Macmillan/McGraw-Hill Scie

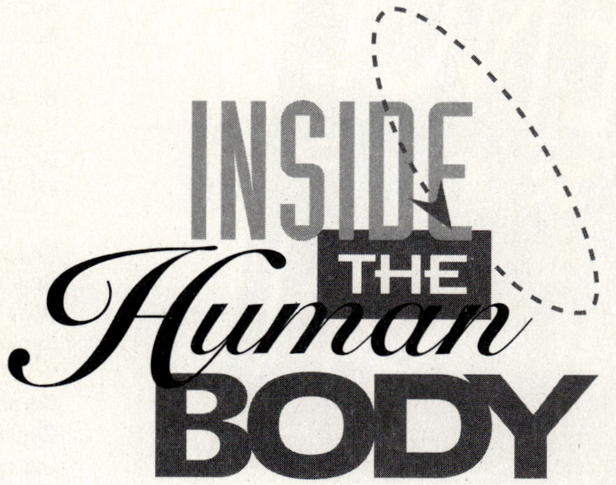

Inside the Human Body

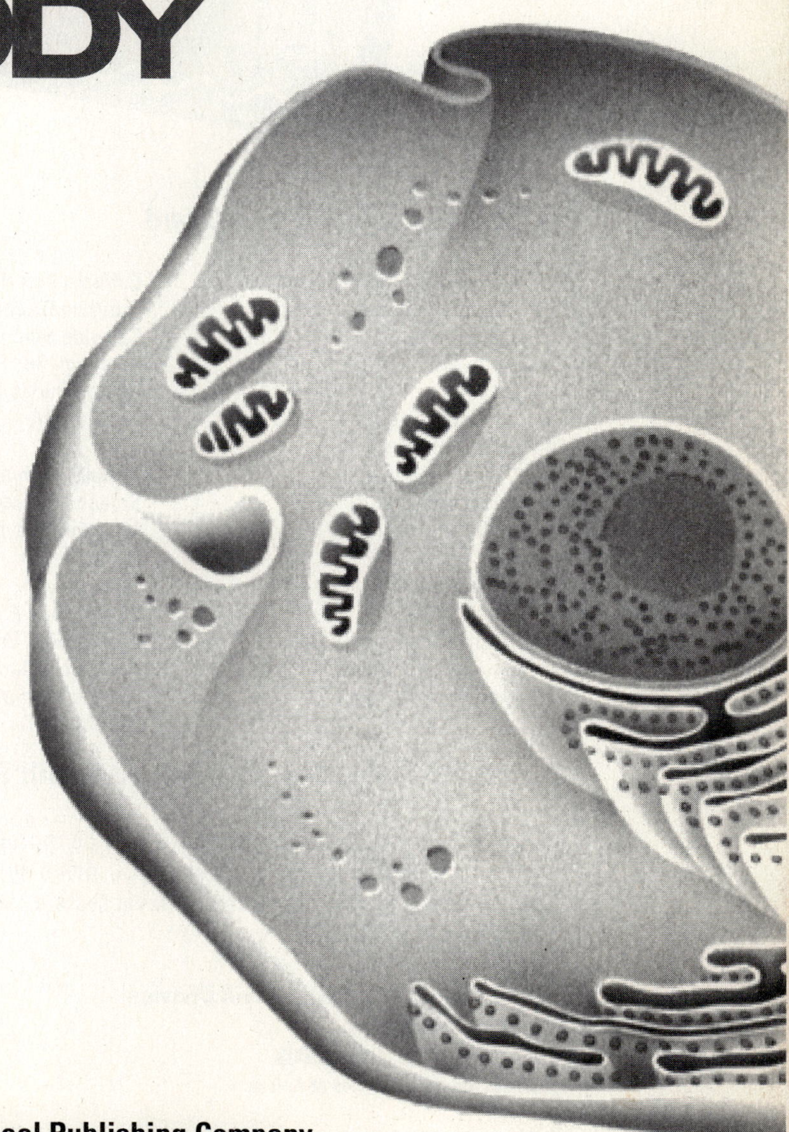

Typical animal cell (not to scale)

AUTHORS

Mary Atwater
The University of Georgia

Prentice Baptiste
University of Houston

Lucy Daniel
Rutherford County Schools

Jay Hackett
University of Northern Colorado

Richard Moyer
University of Michigan, Dearborn

Carol Takemoto
Los Angeles Unified School District

Nancy Wilson-Mathews
Sacramento Unified School District

Macmillan/McGraw-Hill School Publishing Company
New York • Columbus

Macmillan/McGraw-Hill School Division
10 Union Square East
New York, New York 10003
Printed in the United States of America

ISBN 0-02-275958-1

2 3 4 5 6 7 8 9 BAW 99 98 97 96 95 94

The immune system at work

Consultant

Nina M. Surawicz, M.D.
Paradise Valley, AZ

Student Activity Testers

Max Berry
Dustin Derga
Trish Gilfilan
Alveria Henderson
Bri Ramirez

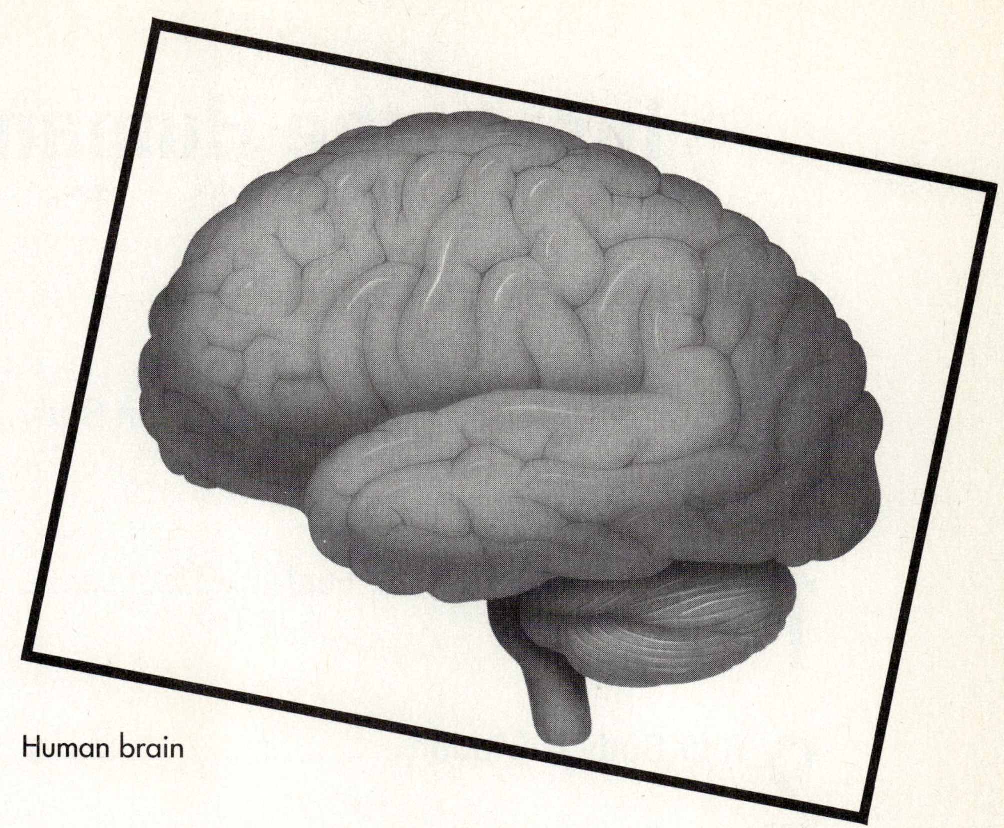

Human brain

Photo Credits:
Cover, ©Meaad Kolyk/Science Photo Library/Photo Researchers, Inc.; **7 (tl),** ©Tom Stack/Tom Stack & Associates, **(tr),** ©E.J. Cable/Tom Stack & Associates, **(bl),** ©H.E. Huxley/Science Photo Library/Photo Researchers, Inc., **(br),** ©D.W. Fawcett/Komuro/Science Source/Photo Researchers, Inc.; **17,** The Bettmann Archive; **20,** ©Studiohio 1993; **33 (l),** ©Studiohio 1993, **(c),** ©Kaz Mori 1992/The Image Bank, **(r),** ©Studiohio 1993; **34,** ©Mary Evans Picture Library/Photo Researchers, Inc.; **37,** ©Visual Horizons/FPG International; **38,** ©Studiohio 1993; **42,** ©Studiohio 1993; **44,** ©Stock Montage, Inc.; **48,** Courtesy of Dr. Bertha Bouroncle; **55 (bl),** ©Adam Hart-Davis/Science Photo Library/ Photo Researchers, Inc., **(bc),** ©1990 A.T. Willett/The Image Bank, **(br),** ©Alfred Pasieka/ Science Photo Library/Photo Researchers, Inc.; **58,** ©CNRI/Science Photo Library/Photo Researchers, Inc.; **59,** ©Studiohio 1993; **62,** ©1990 Howard Sochurek/The Stock Market; **65,** File; **71,** ©Luis Castaneda/The Image Bank; **74 (r),** ©David Krasnor/Photo Researchers, Inc., **(b),** Aaron Haupt/File; **75,** ©Vladimir Lange/The Image Bank; **78,** ©Studiohio 1993; **87,** ©Gail Meese/Meese Photo Research; **88,** ©Howard Sochurek/ The Stock Market; **90,** ©Vladimir Pcholkin 1991/FPG International; **91,** ©Keith Wood/ Tony Stone Worldwide.

Illustration Credits:
6, 16, 18, 24, 25, 31, 46, 57 right, 70, 80, 84, 86 right and top left, John Edwards; **9, 23, 61, 77,** Stephanie Pershing; **14, 40, 41, 43, 45, 54, 56, 68, 69, 72, 81, 82, 83, 85,** Keith Kasnot; **27, 28, 32, 47, 48, 51,** Spencer Phippen; **29, 57 left, 58, 86 lower left, border art,** Scott Leick

Inside the Human Body

Lessons **Themes**

Activities!

EXPLORE

TRY THIS

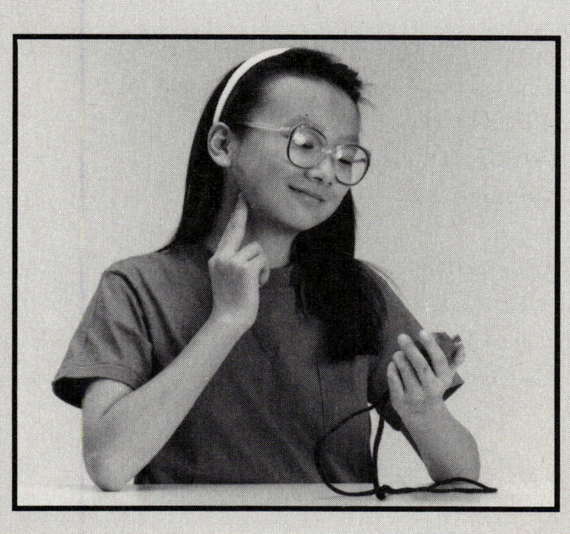

Features

Departments

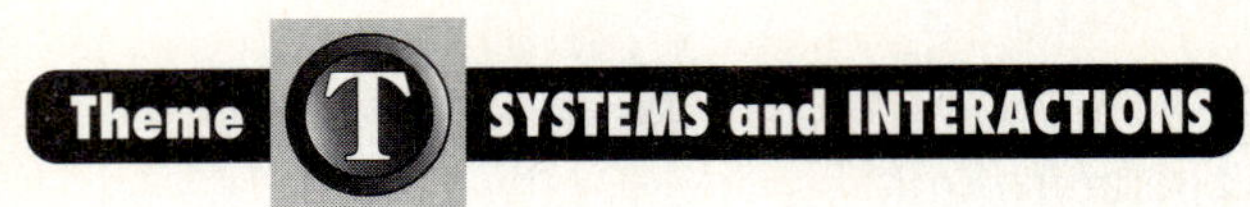

Inside the Human Body

Did you know that your body is made of trillions of cells, all of which came from one cell? And that the cells in your body have hundreds if not thousands of functions, all of which stem from that original cell? In fact, cells are the building blocks of human bodies.

Label this diagram of a cell using the cell parts listed below. Draw a line to connect each cell part in the list with its function at the right. To start, label only those cell parts that you already know. For the parts you do not know the name or the function of, work with several classmates to match a possible name to the cell part. Then pick a possible cell function from the list below.

Cell Parts	**Cell part functions**
Cell membrane	Fluid that allows transportation of materials within the cell
Chromosomes	Barrier that allows only certain things into and out of the cell
Cytoplasm	Cell parts that carry the genetic code of the organism
Mitochondria	Part of the cell within which the chromosomes are found
Nucleus	Part of the cell in which proteins are made
Ribosomes	Cell parts responsible for changing food and oxygen into energy

In living things made of one cell, that one cell has to carry out all life activities. Cells in a many-celled organism, such as a human, do not work alone. Each cell depends on other cells to carry out its functions. This helps the whole organism stay alive.

Cells are specialized according to their functions within the body. Even though a cell may have most or all of the parts of the typical cell above, it may perform a different function from another similar cell located in another part of the body. The four types of cells below, for example, all function differently.

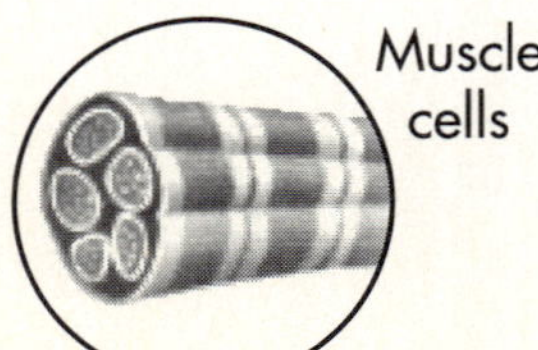
Muscle cells

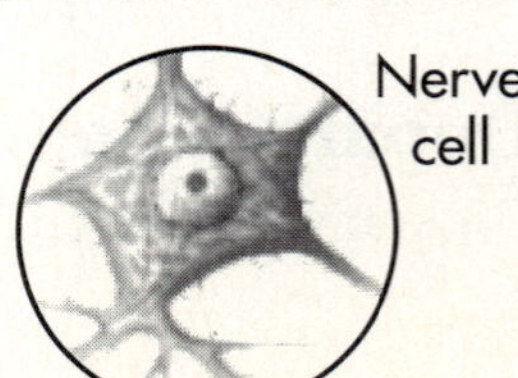
Nerve cell

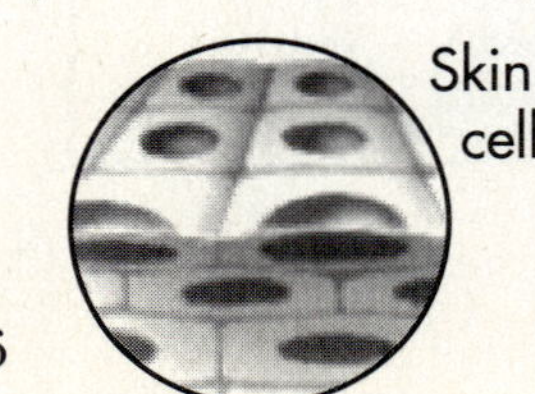
Skin cells

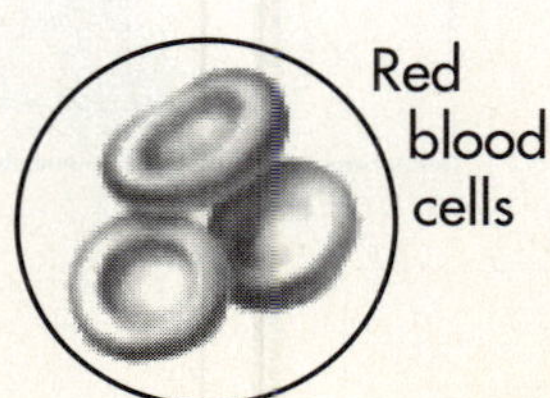
Red blood cells

Cells are the smallest units that carry out the activities of life. In many-celled organisms, cells are organized into tissues. Tissues are groups of similar cells that work together to perform the same function.

You may be surprised to learn that there are only four different kinds of tissue—epithelial (ep′ ə thē′ lē əl), connective, muscle, and nervous.

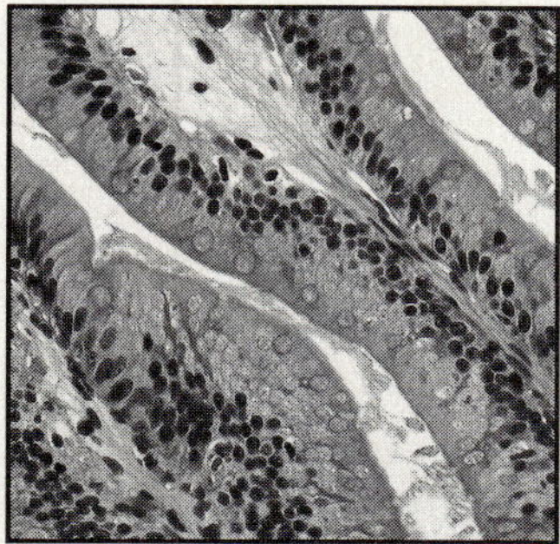

Epithelial tissue is found wherever parts of the body need a covering or a lining. For instance, the lining of the stomach is made of epithelial tissue.

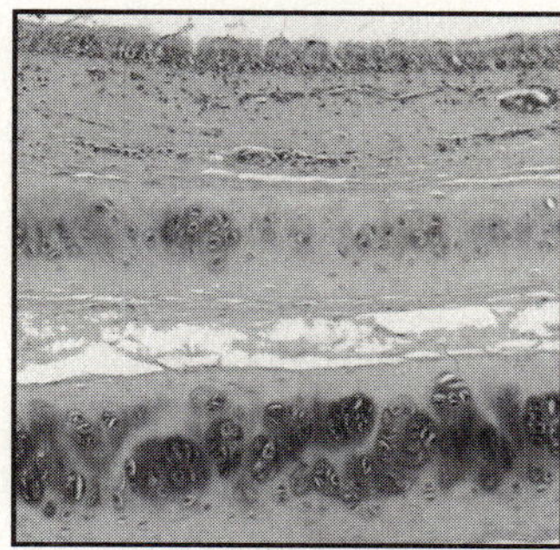

Connective tissue includes bone, cartilage, blood, and fat. These tissues serve a variety of functions in the body including insulation, protection, and support.

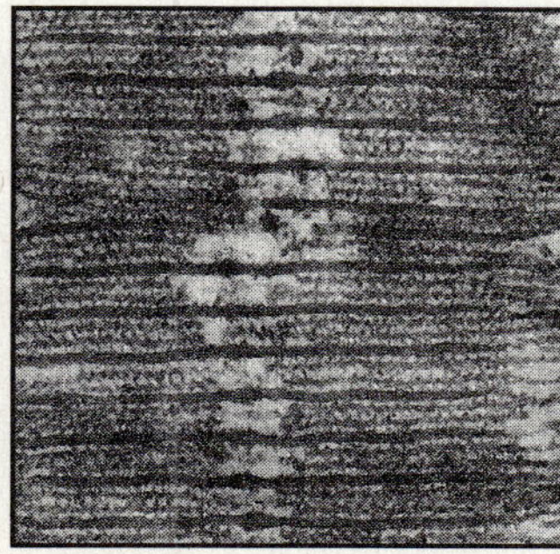

Muscle tissue makes up the muscles of the body.

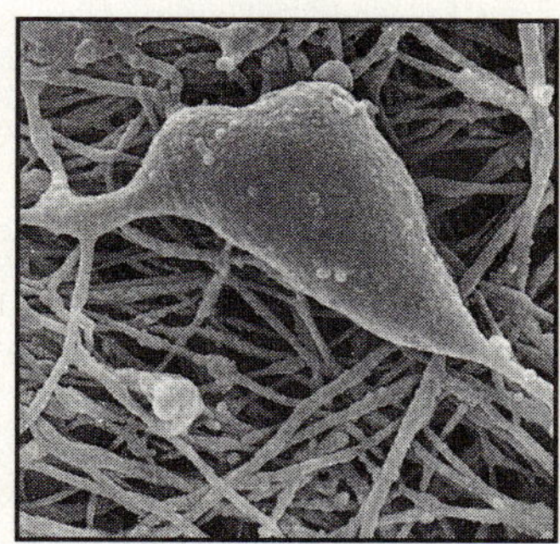

Nervous tissue consists of nerves throughout the body, including those of the brain and spinal cord.

Different kinds of tissues work together to form an organ. An organ is a group of tissues working together to perform a certain function. An example of an organ is the skin, which is composed of epithelial tissue, connective tissue, and some muscle and nervous tissue. Each separate bone of the body is an organ made up of several different types of connective tissue.

A group of organs working together to perform a certain function is a system. Body systems work together to keep the body functioning properly and to help the organism stay alive. As you read through this unit, you'll learn how body systems operate and what functions they fulfill within the body.

Turning Food Into Body Fuel

Congratulations! You are the proud owner of a complete human digestive system. When properly used and maintained, your digestive system can provide you with years of energy for your daily functions. The trillions of cells that make up your body need a source of nourishment to remain healthy and to keep you physically and mentally active. It's the job of your digestive system to transform food into a form your cells can use for energy.

Minds On! Draw a diagram of your digestive system on a separate sheet of paper. Label the major parts . Do you know what each part does? If so, write a brief description of the function of the part next to it in the diagram. If you had to compare your digestive system to something else, to what would you compare it? Share your answers with others in your class. ●

Nutrient Detective

Different types of food can be separated in many different ways; for example, by tastes, by types, or by colors. In this activity you'll learn another possible way to separate food into types—a way that becomes very important as you continue to study this lesson.

What You Need

iodine solution in dropper bottle
Biuret solution
clear containers
rubber gloves
dropper
brown or white paper
a variety of foods to test
plastic wrap
safety goggles
rubber band

What To Do

1 ***Safety Tip:*** Wear your goggles and rubber gloves throughout steps 1-4 of this activity. Place one food into each container. Add about 1 dropperful or 5 mL of iodine solution to each container. Record what happens to the color of the food. ***Safety Tip:*** Iodine is poisonous and can stain clothes and skin. Do not get any on your clothes and skin, and do not breathe the fumes. Dispose of the mixture in the container as your teacher instructs you.

2 Repeat step 1 until all foods are tested.

3 Place one food into each container. Add 1 dropperful or 5 mL of Biuret solution to each. Carefully seal the container using the plastic wrap and rubber band. Gently swirl or shake the solution to mix. Record your observations. ***Safety Tip:*** Biuret solution can harm your skin. Make certain that you do not allow it to come into contact with your skin. Dispose of the mixture in the containers as your teacher instructs you.

4 Repeat step 3 until all foods are tested.

5 Test each food by rubbing a thin layer of it on clean paper. Let the mark dry if necessary. Hold the paper up to the light. Record your observations.

See the ***Safety Tips*** in steps 1 and 3.

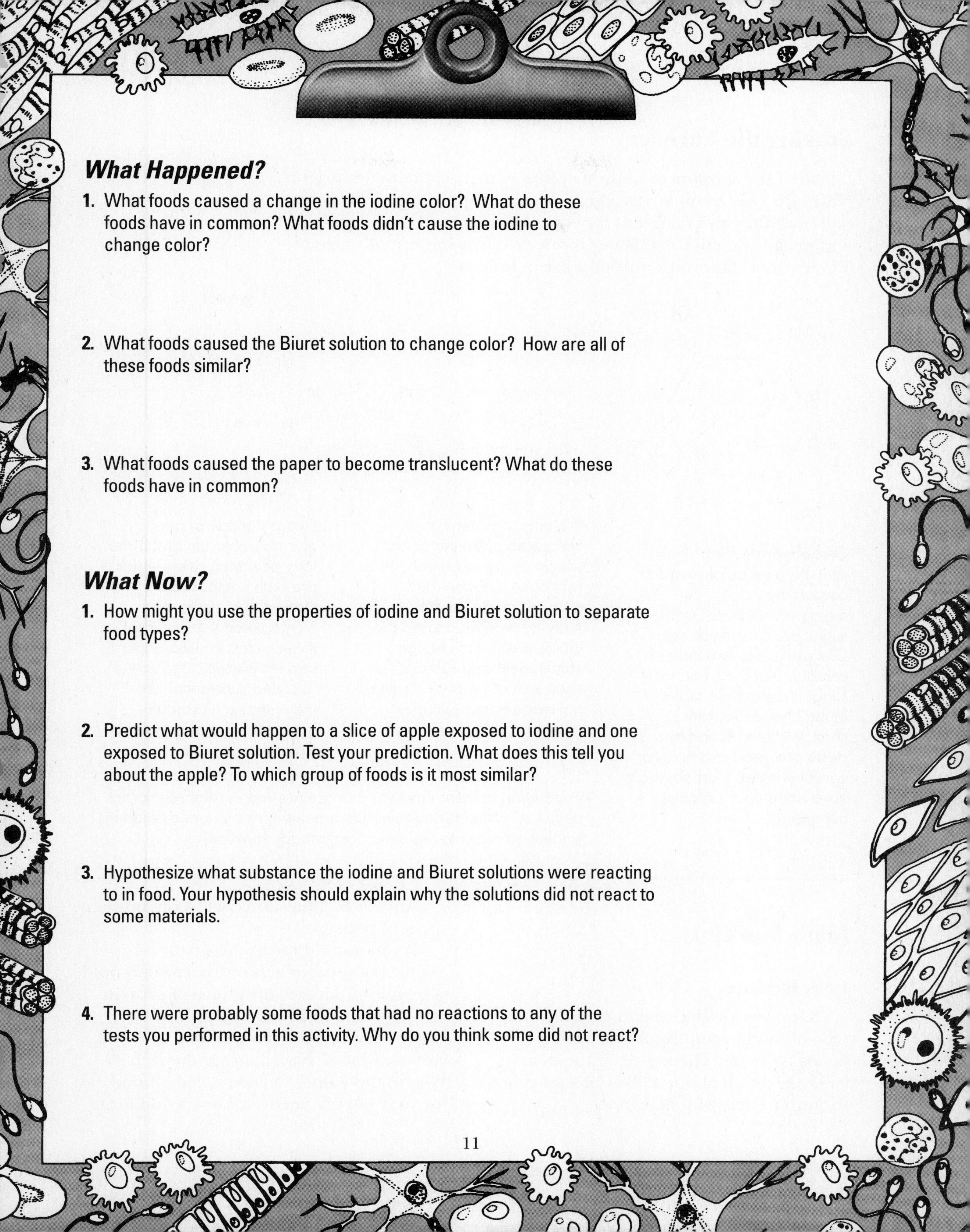

What Happened?

1. What foods caused a change in the iodine color? What do these foods have in common? What foods didn't cause the iodine to change color?

2. What foods caused the Biuret solution to change color? How are all of these foods similar?

3. What foods caused the paper to become translucent? What do these foods have in common?

What Now?

1. How might you use the properties of iodine and Biuret solution to separate food types?

2. Predict what would happen to a slice of apple exposed to iodine and one exposed to Biuret solution. Test your prediction. What does this tell you about the apple? To which group of foods is it most similar?

3. Hypothesize what substance the iodine and Biuret solutions were reacting to in food. Your hypothesis should explain why the solutions did not react to some materials.

4. There were probably some foods that had no reactions to any of the tests you performed in this activity. Why do you think some did not react?

Stoking the Furnace

All of the substances in the Explore Activity contained nutrients. Nutrients (nü′ trē ənts) are chemicals that supply energy for your daily activities and materials for your body's growth, maintenance, and proper functioning. Water is one critically important nutrient. Three other important nutrients are as follows.

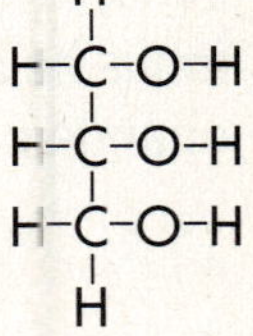

Glycerol

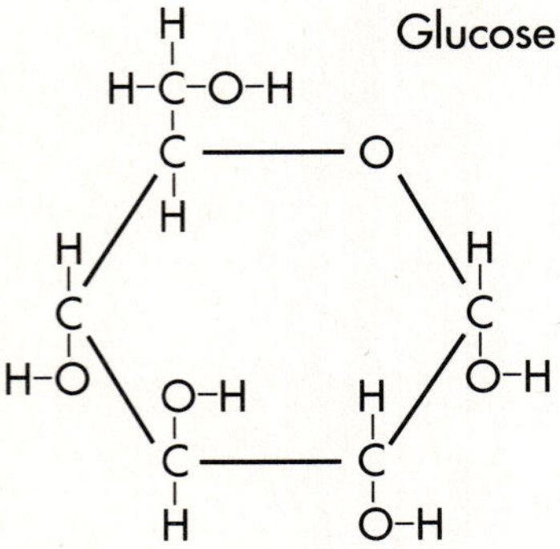

Glucose

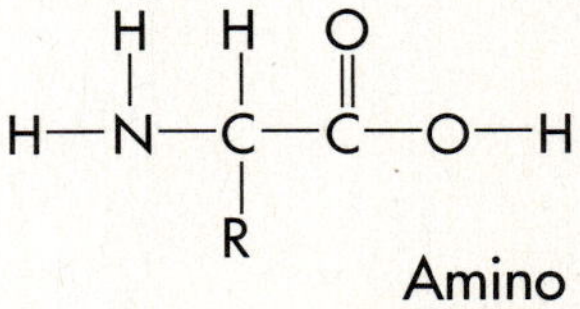

Amino acids

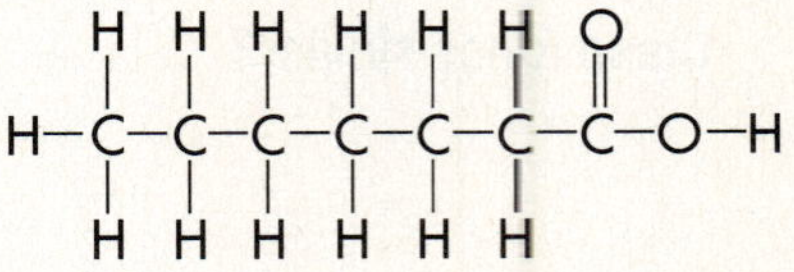

Fatty acid

Carbohydrates (kär′ bō hī′ drāts) are made of atoms of carbon, hydrogen, and oxygen. Both starch and sugar are carbohydrates. Glucose is one example of a sugar. Starches, but not sugars, caused the iodine in the Explore Activity to change colors. Bread and pasta are good examples of carbohydrates. Carbohydrates have about four Calories per gram.

Proteins (prō′ tēnz) are composed of amino acids, which consist of carbon, hydrogen, oxygen, and nitrogen atoms. In the Explore Activity, proteins caused the Biuret solution to change colors. Beef and fish are examples of proteins. Proteins have about four Calories per gram.

Fats are made of one glycerol molecule and three fatty acid molecules, which are both composed of carbon, hydrogen, and oxygen atoms. In the Explore Activity, fats caused paper to turn translucent. Egg yolks, oils, and substances like mayonnaise all contain fats. Fats have about nine Calories per gram.

In addition to these nutrients that are needed in relatively large amounts, vitamins and minerals are nutrients that are needed in smaller amounts to maintain proper body functioning.

Math Link

Less Is More?

Suppose you decided to eat only one type of food to supply all your energy needs for a day. The average human at your age needs about 4,000 Calories a day to supply energy for activities.

Answer the questions below on another sheet of paper.

If you wanted to eat the smallest amount of food, which nutrient type would you choose? How many grams of protein would you need to eat to supply your energy needs? How many grams of carbohydrates? How many grams of fats? Why would eating only one type of food for your energy needs not be a good idea?

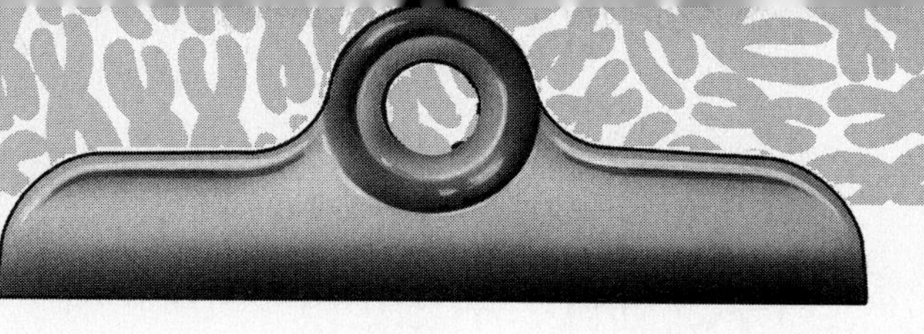

TRY THIS **Activity!**

Balancing Act

Putting together a balanced diet is not as easy as it seems. Sometimes the things we like to eat best are the ones we should be eating the least. Do the activity to try to make a menu for one day.

What You Need

Food Guide Pyramid
Vitamins and Minerals Chart
11" x 17" piece of oaktag/posterboard
colored markers

What To Do

Carefully examine the diagrams your teacher will provide. Use the information on the diagrams to plan a menu, including any snacks, for a single day. Consider how you will get all of the allowances of different food groups, vitamins, and minerals. On another sheet of paper, write a menu for one day that would include the proper balance of foods from each food group. Check to see that as many vitamins and minerals as possible are present. For comparison, write down what you ate yesterday.

Once you have planned the menu, make a poster on your posterboard that illustrates your suggested meals. Of which food group do you consume the most? Of which do you eat the least? Which food group do you like the best? The least? Share your posters with other class members and see if you can come up with a week or more of healthy meals that include all the vitamins and minerals. Compare your normal menu with your balanced menu. How well do they match? How easy was your balancing act?

Through the Lips and Over the Tongue

How does your body obtain nutrients? Start your study of the digestive system by writing the names and descriptions of the functions of the digestive organs in the appropriate blanks.

Part: ____________________
Function: ____________________
____
____

Part: ____________________
Function: ____________________
____
____

Pharynx

Part: ____________________
Function: ____________________
____
____

Part: ____________________
Function: ____________________
____
____

Part: ____________________
Function: stores enzymes from the liver
____

Part: ____________________
Function: absorbs water
____
____

Part: ____________________
Function: secretes enzymes that help digest carbohydrates, fats, and proteins

Part: ____________________
Function: ____________________
____
____

Appendix

Part: ____________________
Function: ____________________
____

Anus

Body Part

Esophagus
Gallbladder
Large intestine
Liver
Mouth
Pancreas
Rectum
Small intestine
Stomach

Function

Collecting area for wastes
Completes digestion of food and absorbs food
Connects the pharynx to the stomach
Food enters the body and is broken down by chewing
Holding place for food, mixes food with enzymes
Secretes enzyme (bile) that helps digest fats

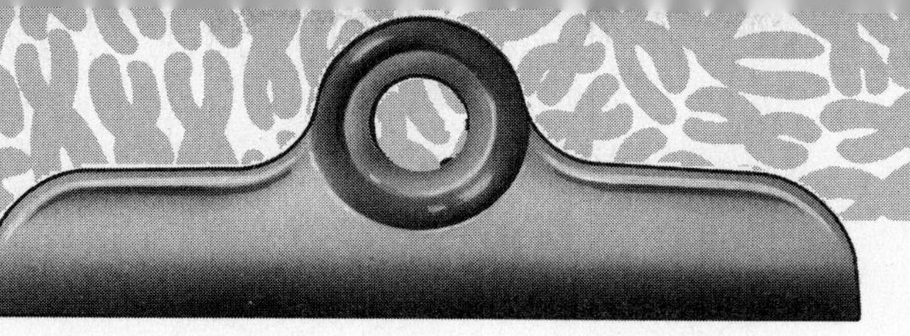

TRY THIS

Activity!

The Bigger, the Better?

Why is it important to break food down into smaller pieces before digestion begins? Do the activity to find out.

What You Need

2 pieces of filter paper
solid piece of colored chalk
2 funnels
2 250-mL beakers
safety goggles
powdered colored chalk
200 mL vinegar
2 plastic jars
scale or balance

Safety Tip: Wear your goggles throughout this activity.

On one side of the balance, place a piece of chalk on a piece of filter paper. On the other side place a piece of filter paper. Place enough chalk powder on the filter paper to balance the piece of chalk. Put the chalk in one jar and the chalk powder in the other. Add 100 mL of vinegar to each jar. Wait 5 minutes.

Fold your filter paper to line each funnel. Place the funnels into beakers. Pour the contents of each jar into a funnel. Rinse the jars and filter papers thoroughly with water. Remove the filter papers from the funnel and allow them to dry. Put the two pieces of filter paper with their contents onto the balance. Record all observations on another sheet of paper, and answer the following questions.

1. Is there a difference between the mass of the two pieces of filter paper and chalk now? If so, which one has greater mass?

2. Suggest a hypothesis to explain what you have observed.

3. Explain how this model helps you understand the role of chewing in digestion.

Digestion begins in the mouth as food is physically broken down by chewing. At the same time, it is mixed with a clear liquid called saliva (sə lī′ və) that contains **enzymes** (en′ zīmz)—chemicals that help speed up the process of digestion. One enzyme in saliva starts to break starches down into simple sugars.

When food is thoroughly chewed, the tongue pushes it against the top of the mouth. This action forces food back into the pharynx (far′ ingks) a short tube leading to the esophagus (i sof′ ə gəs). The esophagus is a tube connecting the pharynx to the stomach.

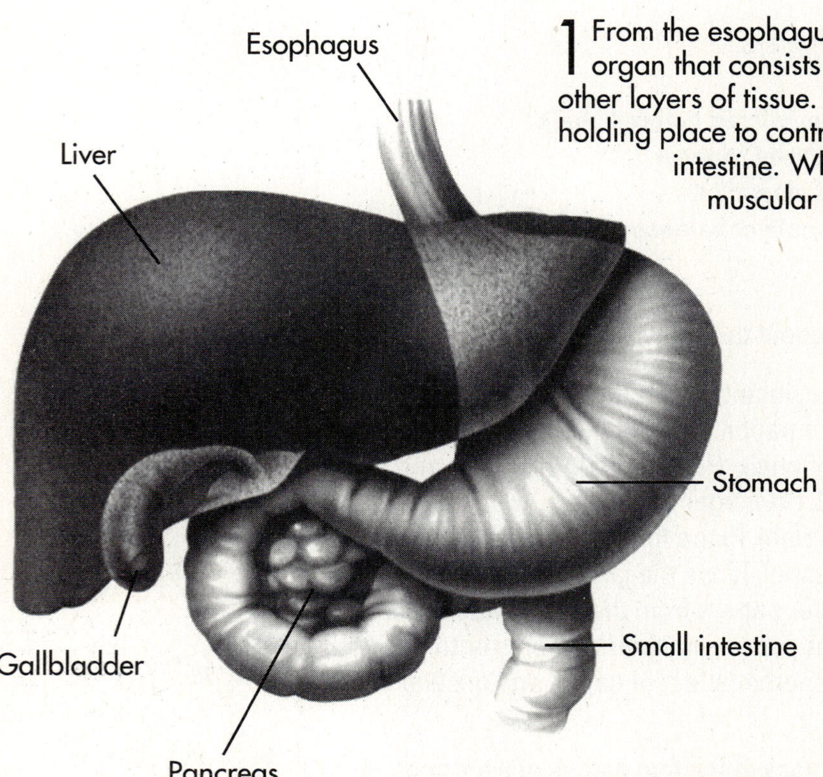

1 From the esophagus food enters the **stomach**, an organ that consists of three layers of muscles and other layers of tissue. The stomach functions as a holding place to control the entry of food into the small intestine. While food is in the stomach, muscular action continues the physical breakdown of food and mixes food with gastric, or stomach, juices. Two important chemicals that the stomach adds to food are hydrochloric acid and an enzyme that helps break down proteins.

The stomach absorbs very little food. However, both alcohol and aspirin can be absorbed through the stomach walls. This is one of the reasons alcohol can act so quickly in the body to impair functioning.

2 From the stomach food is passed into the **small intestine** where digestion is completed. Here food is mixed with juices from the gallbladder and pancreas. The gallbladder stores bile from the liver. Bile causes large fat droplets to break up into smaller droplets.

3 The pancreas produces important chemicals for digestion. One enzyme helps break down starch, one helps break down proteins, and one helps break down fats.

Social Studies Link

Looking Into Digestion

Much of our understanding of the digestive system comes from advances made in scientific techniques. However, careful observations and experiments are also extremely important. One of the milestones in understanding digestion occurred in 1822 when a Michigan man named Alexis St. Martin accidentally shot himself in the stomach with a musket.

His doctor, William Beaumont, helped St. Martin recover from the injury, which at that time normally would have been fatal. One of the odd results of the accident was that St. Martin developed an open hole in his stomach. Nothing Beaumont did helped to close the hole. When the wound was closed with a bandage, digestion could occur as usual.

Beaumont offered St. Martin free room and board in exchange for St. Martin's participation in a series of safe, painless experiments. Each day Beaumont would tie a different kind of food to a silken thread and lower it into St. Martin's stomach. They would close the hole, and hourly Beaumont would look at the food to see what was happening to it.

Through an unfortunate, nearly fatal accident, medical science was able to get another glimpse of how the digestive system worked. What do you think Beaumont saw? Pretend you are Beaumont's assistant during one of these experiments. On a separate sheet of paper, record the results of Beaumont's experiments for the day. How were different types of foods affected? What do you think he saw when he put in such foods as steak, bread, animal fat, boiled eggs, or crackers? Tell what Beaumont did that day, and what his observations were. Also record your own reactions. How would such an experiment help you understand digestion?

Sketch showing Dr. Beaumont and St. Martin

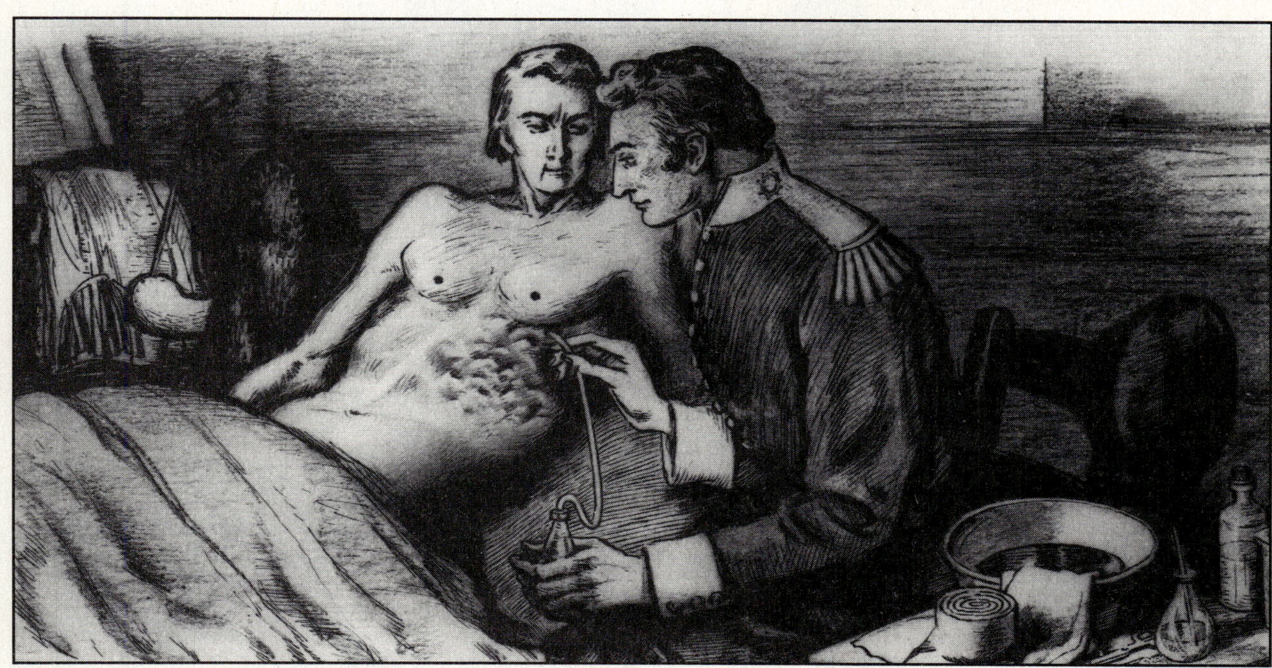

Completing Digestion

After food has gone through the stomach and has been mixed with the chemicals from the pancreas and liver, it is almost ready to be used by the body.

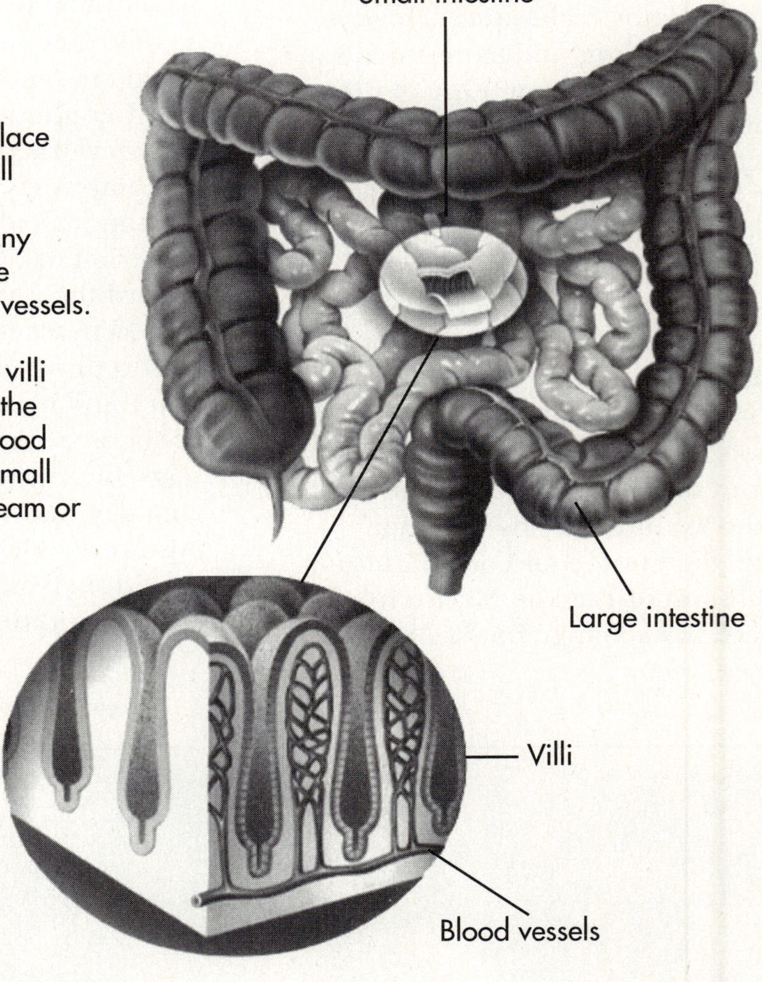

Small intestine

Large intestine

Villi

Blood vessels

The last steps of digestion take place near cells in the walls of the small intestine. The wall of the small intestine is lined with a great many tiny folds called **villi** (vil′ ī). These folds are filled with many blood vessels. On the membranes of cells that line the intestinal side of the villi are enzymes that help complete the digestion of food. The digested food passes through the walls of the small intestine and enters the bloodstream or another transport system, the lymphatic system. The digested food is carried throughout the body to provide the nutrients and energy necessary for life functions.

From the small intestine undigested food, or waste, enters the large intestine. The **large intestine** is so called because of its large diameter. The large intestine does not absorb food, but the walls of the large intestine do absorb water from the waste that passes through it. This water is used in many different body processes.

From the large intestine, waste material that the body cannot use moves into the rectum and is finally expelled from the body through the anus. This waste material is called feces (fē′ sēz).

Troubleshooting Guide to Your Digestion

Guide to Digestion- and Nutrition-related Diseases and Disorders

Disease/Disorder	Cause	Effect on the Body	Treatment/Avoidance
Dental caries (tooth decay)	Consuming food with too much sugar, infrequent brushing and flossing	Decay and eventual loss of teeth	Frequent brushing and flossing, controlling sugar consumption
Gallstones	Buildup of mineral matter in the gallbladder	Acute pain and nausea upon eating some foods	Break up stones or remove gallbladder
Indigestion	Irritation of the stomach lining	Improper eating, consuming alcohol or aspirin	Eat regularly, avoid substances that cause stomach-lining irritation
Constipation	Not enough fiber and water in the diet	Sluggishness, discomfort	Eat a balanced diet with lots of high-fiber foods
Kwashiorkor (kwä′ shē ôr kôr)	Severe malnutrition	Swelling of the limbs and extremities, dangerous thinness	Food, particularly protein, in a balanced diet

Review the diseases and disorders included in this chart. Brainstorm with your group to list other diseases and disorders. Choose one to research. Include information on cause, effect, and treatment of the disorder or disease. Share your information with the class and complete the chart.

Hospital Dietitian

Fatima Bellows is a vital member of a team including nurses, doctors, family, and friends that helps a patient in the hospital recover from surgery. Fatima is a hospital dietitian.

Working closely with doctors, Fatima plans the menus for patients throughout the hospital. For a patient who has had surgery, it's important to know what kind of surgery, how long ago, and what sort of progress the patient is making toward recovery. She also counsels the patient on the importance of continuing to follow the suggested diet once back at home. Good nutrition is essential to the patient's complete recovery.

In order to work as a hospital dietitian, Fatima had to earn a four-year degree in nutrition. She also had to be licensed in the state in which she works.

Try your hand at Ms. Bellows' job. Your friend Agnes went to the doctor, who recommended that she cut down on fat in her diet. Write a menu for one day of Agnes' diet on a separate sheet of paper.

Ms. Bellows carefully plans menus for patients.

Literature Link

Food, Nutrition, and You

What special nutritional needs do you have? If you're an athlete, should you try to eat more vitamins, minerals, and proteins than a nonathlete would need? If you're a vegetarian, are you getting the right nutrients to build strong bones and teeth? Do girls your age have different nutritional needs than boys?

Learn the answers to these questions and hundreds more by reading *Food, Nutrition, and You* by Linda Peavy and Ursula Smith. Then, with several classmates, choose one nutritional fact that surprised you from the book and decide how to teach it to the rest of the school. For instance, you might make a poster or a banner to hang near the cafeteria.

Sum It Up

Your body needs food as its source of energy and materials for tissues and tissue repair. Proper nutrition is an important part of keeping the digestive system and your entire body functioning as it should. Proper nutrition includes a balanced diet of proteins, carbohydrates, fats, vitamins, minerals, and water.

The organs of your digestive system work together to break food down into a form that can be used by cells in your body. This process of converting food into a source of fuel and materials for body repair is necessary for all life processes.

Using Vocabulary

Draw and label a map of the digestive system. Write a travel guide to your digestive system using the words in the vocabulary list to explain the process of digestion. Make sure that someone who has not read this lesson can understand your travel guide.

carbohydrates	large intestine	stomach
enzymes	proteins	villi
fats	small intestine	

Critical Thinking

1. Explain the difference between physical breakdown of food and chemical breakdown of food. Which is more important in digestion? Why?

2. What might happen to young children who don't get enough protein in their diets?

3. How might having your stomach removed affect your ability to digest starches?

4. How are proteins and carbohydrates different?

5. If a customer came into your grocery store and asked for a high-fiber food with lots of vitamin A, what foods would you recommend to him or her?

Your Body in Motion

Motion—playing basketball or soccer, running, climbing stairs, or fidgeting in your seat to get more comfortable—all of these actions are yours courtesy of your muscular and skeletal systems. Yes, this complete wide range of movements is possible because of two systems that work together to bring motion to your life.

How does this happen? How can you assure that your motion systems will remain in peak working order? How do these systems interact to use the fuel provided by your digestive system to keep you in motion? In this lesson you'll find out.

Minds On! With your group, take five minutes and develop a list of all the motions you go through during a regular school day. Make your list as complete as possible. How do you think the muscular system and the skeletal system make these motions possible? Record your answers on another sheet of paper and be sure to leave plenty of room for additions and explanations as you go through the lesson. ●

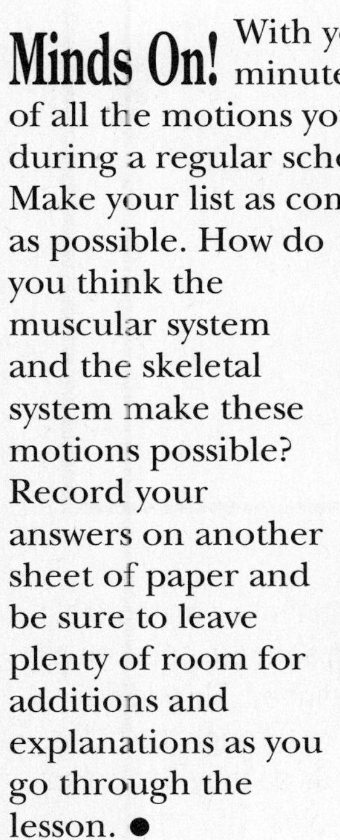

How Do Muscles and Bones Produce Motion?

What roles do muscles and bones play in motion? How do muscles produce motion? Would it be possible to move in the same way without bones? In this activity, you will make a model of some muscles and bones that will help you answer these questions.

What You Need

3 sheets posterboard (about 30 cm × 15 cm)
masking tape
chenille wire
2 long balloons
black marker
hole punch
meter tape

What To Do

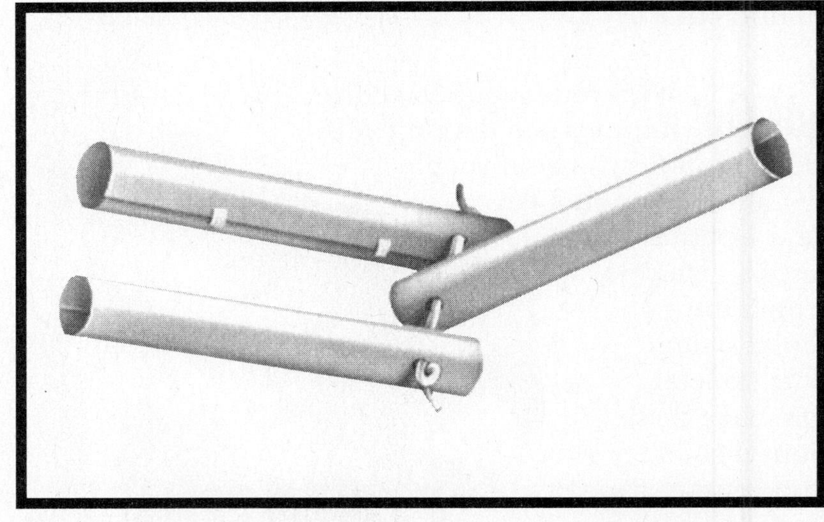

1 Carefully roll the three pieces of posterboard into long tubes. Tape the tubes so that they hold their shape.

2 Use the hole punch to punch two holes, one opposite the other in a straight line, near one end of each tube.

3 Place the tubes on the table making sure the holes on all three are aligned. Thread the chenille wire through the holes. Curl the ends so they don't slip out of the holes. The chenille wire should now be holding your tubes together.

4 Tape two of the tubes together as shown in the illustration. ▶

5 Write the letter A on one balloon and the letter B on the other. Stand the middle tube up to form an L-shape. Inflate each balloon to a length of 20 cm and tie them to hold the air in. Tape balloon A near the top of the single tube facing toward the two tubes. Tape balloon B on the opposite side of the single tube. Tape balloons A and B to the two tubes as shown.

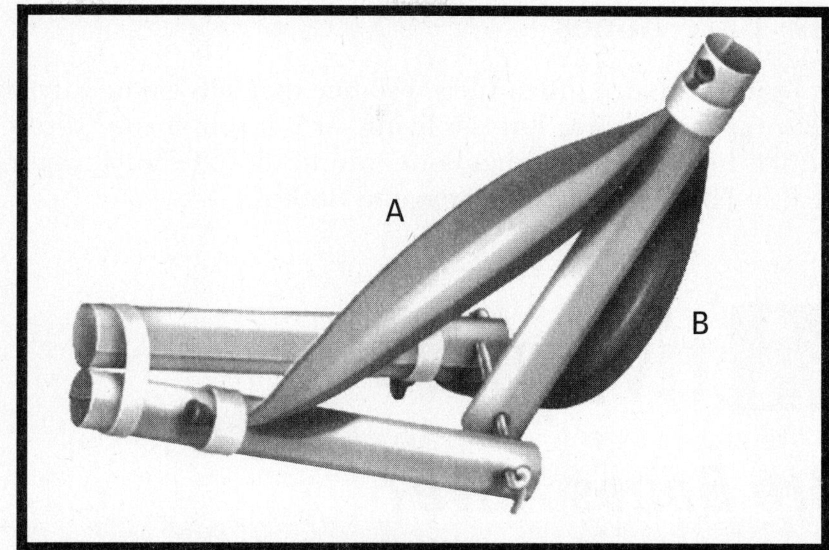

6 Move the tubes to try to straighten them out. Observe what happens to the balloons. Record your observations. Experiment with the balloon and tube model, moving it in a variety of ways to see what happens. Record your observations and answer the questions below on another piece of paper.

What Happened?

1. What body part did you construct a model of? How can you tell? What function did each part serve?
2. What happened to each balloon when you bent the tubes toward each other? When you straightened them out?
3. Which balloons were doing the work in each case? How could you tell?

What Now?

1. If this were living material, how could the movement of the balloons cause the tubes to move?
2. Try to develop a way to move the tubes without causing the balloons to contract or extend. Were you able to? What does this tell you about movement in a real body part?
3. How would the motions differ if there were no cardboard tubes to attach the balloons to?

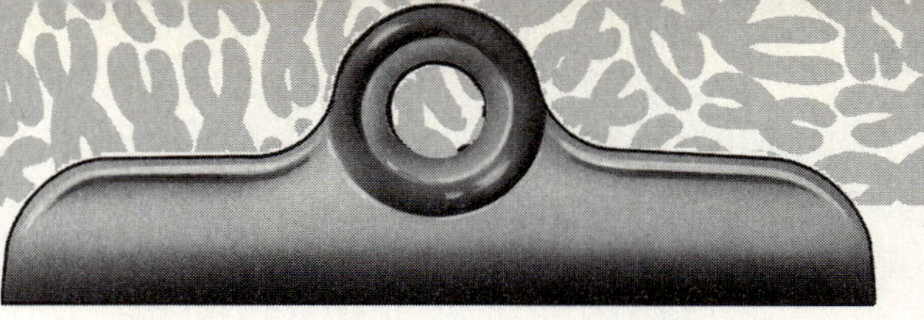

The Bare Bones

Bone is a hard, inflexible substance that acts as the internal support for the body. But is it living, or is it non-living? If you could extract the minerals from a bone, would there be anything left? Do the Try This Activity on this page to find out.

TRY THIS **Activity!**

Are Bones Alive?

Are bones a living part of a body, or are they simply a structural support made of solid minerals, like a rock? How can you tell if they are alive?

What You Need

cleaned chicken leg bone	**250 mL vinegar**
empty plastic container with lid	**safety goggles**

Place the bone in the container. Fill the container with enough vinegar to cover the bone completely. Cover the container and leave it in a cool place for two days. Pour the vinegar off the bone. Rinse the bone.

1. Does the bone feel the same as it did before you put it in vinegar?

2. What did the vinegar do to the bone?

3. If the bone had been solid mineral, would there be anything left of it?

4. Formulate a hypothesis telling why the bone did not entirely dissolve.

In the activity, you saw that after you removed much of the mineral material from bone there was still bone left. What was left was the once-living tissue of which bone is made. To understand more about bone and bone structures, study the illustration on this page.

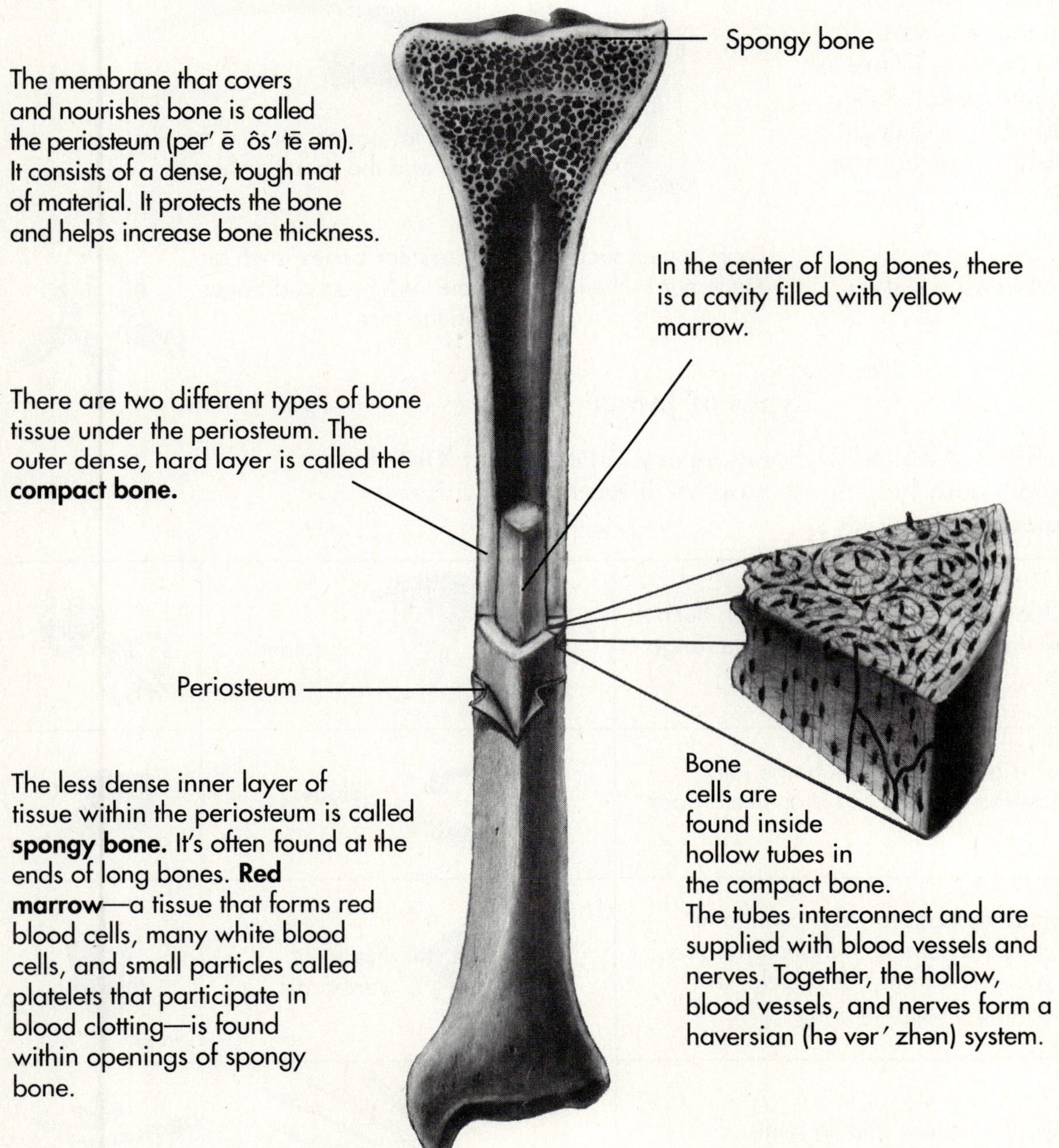

Types of Bones

Bones are grouped into different types depending on their shape.

Minds On! On another sheet of paper, name as many different types of bones as you can and write a brief description of how each functions in your body. ●

Short bones, such as those in the fingers and toes

Flat bones, such as the ribs and the shoulder blades

Illustrations not to scale

Long bones, such as the femur in the leg

Irregular bones, such as the vertebrae and bones in the face

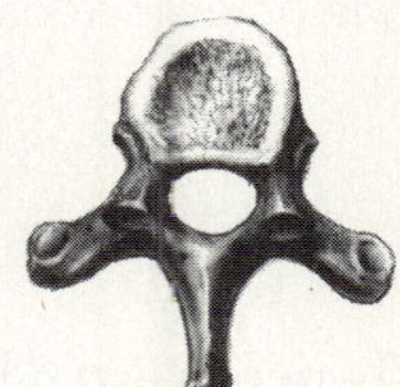

Types of Joints

The point at which two bones meet is called a joint. Different joints in the body help bones move in different ways with respect to each other.

The shoulder is a ball-and-socket joint. Such a joint allows for a complex and large range of motion.	Shoulder	
The elbow is a hinge joint. A hinge joint allows movement in one direction like a door hinge.	Elbow	
Your topmost vertebra forms a pivot joint with the vertebra beneath it that allows you to move your head from side to side, as though pivoting around an axis.	Neck vertebrae	
Joints in your wrist are gliding joints. Movement in gliding joints is quite complex.	Wrist	

Literature Link

You've read about the skeleton and have a good idea of its importance in the human body. What would it be like to be without a skeleton? What might some of the problems be?

Writers often explore imaginary worlds in which the impossible occurs. One writer, Ray Bradbury, has even considered the importance of the skeleton. Read Bradbury's story "Skeleton," which is about a man who becomes very aware of how his skeleton helps him function.

Discuss the story with your group. When you're finished, work with your teacher to choose one of the activities below:

- Choose three or four main events in the story and illustrate them.

- Continue the story. Tell about the next day in the life of Mr. Harris. What does he do? How does he eat?

- Now that you've read one story about skeletons, write your own short story or poem about the importance of the skeletal system. Or compose a song on the same subject.

Muscles, the Body's Movers and Shakers

Bones provide support and perform other important functions in the body. However, as you saw in the Explore Activity, bones do not move by themselves. They must be moved. You are probably aware that muscles help move bones. Do the Try This Activity to learn about the several different kinds of muscles.

TRY THIS Activity!

Muscular Types

Different types of muscles perform different functions in the body. What do the different types of muscles look like?

What You Need

microscope, paper, pen or pencil
prepared slides labeled as follows: cardiac muscle, smooth muscle, skeletal (striated) muscle

Take one of the three slides and look at it. On a separate sheet of paper, draw and label a diagram of what you see through the microscope. Repeat for the other two slides. Be careful not to confuse your drawings and your labels.

1. How did the slides differ from one another? What features in each slide were similar or the same?

2. From the name, where do you suppose cardiac muscle might be found?

3. What do you think might be one of the functions of skeletal muscle?

4. Which types of muscle are more similar–smooth and skeletal, skeletal and cardiac, or cardiac and smooth? How might this similarity relate to their functions in the body?

5. What do you think smooth muscle might do in the body?

Cellular Powerhouses

You know that motion requires energy. How do the muscles and bones get the energy they need for motion? You know your digestive system provides fuel for your body, but how does your body use this fuel?

The fuel your digestive system provides is carried by the blood to all the cells of your body. Within cells the fuel provided by your digestive system is processed and the resulting energy is stored in chemical bonds. The process of converting fuel to energy within the cell is called **respiration**. Much of the processing takes place within the mitochondria. The mitochondria are rightly called "The Powerhouses of the Cell."

As fuel is converted to energy, oxygen is used in the cell and carbon dioxide and water are given off. The energy is stored in chemical bonds. When the cell needs energy, chemicals in the cell break the energy-storing bonds. Energy is released for use in your daily activities and body processes.

Minds On! Use the diagram of a cell below and a separate piece of paper to draw a model of the process you've just read about. Indicate the materials that must enter the cell, those that are used in the cell, what is produced, and what is removed from the cell as waste products. ●

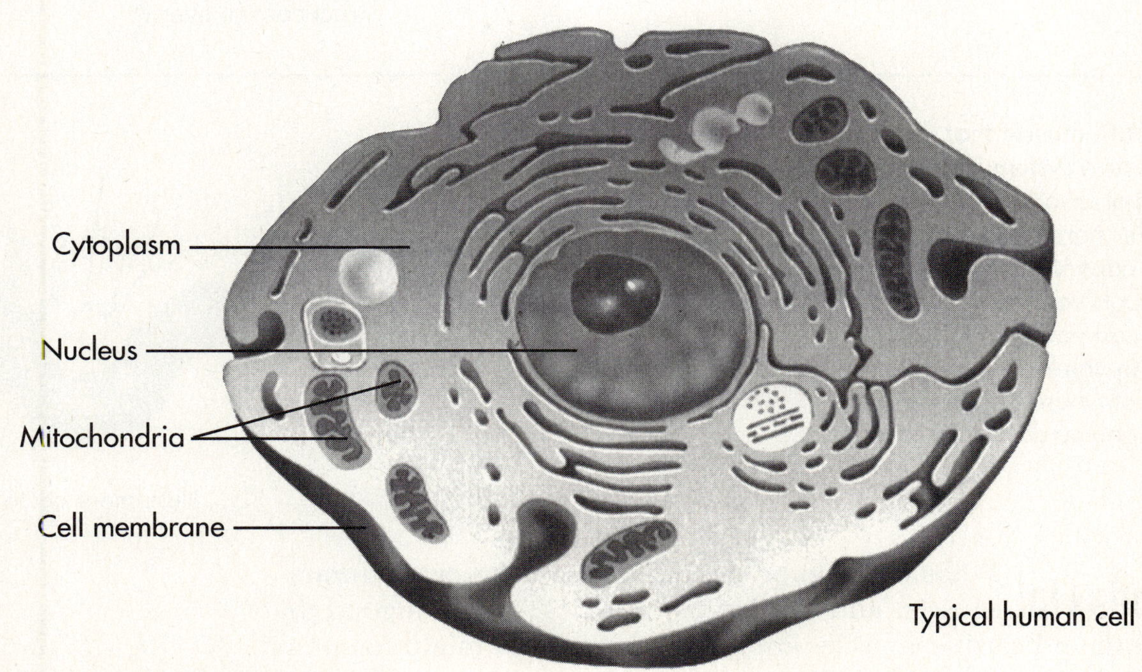

Typical human cell

Getting Into Motion

What types of motion is the body capable of? You know that you can determine how quickly you move. You can choose to walk, jog, or run. This is because the muscles that control those kinds of movements, **skeletal muscles**, are under your control. They are called voluntary muscles.

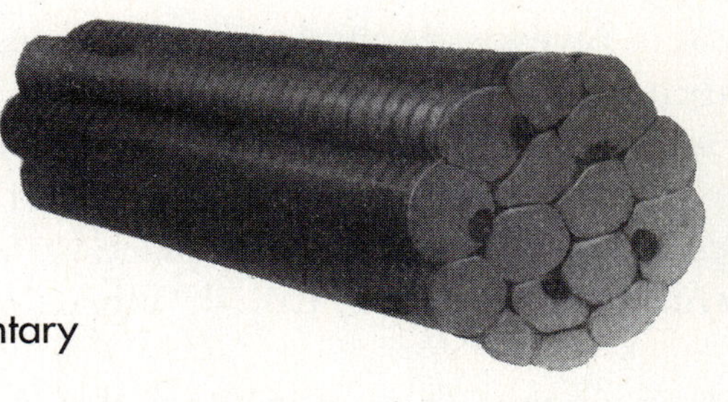

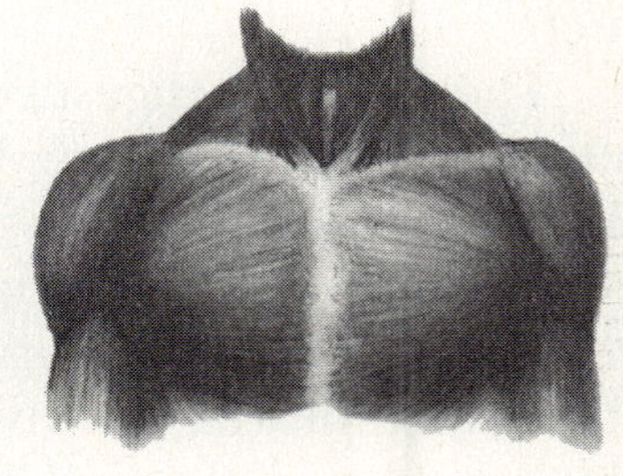

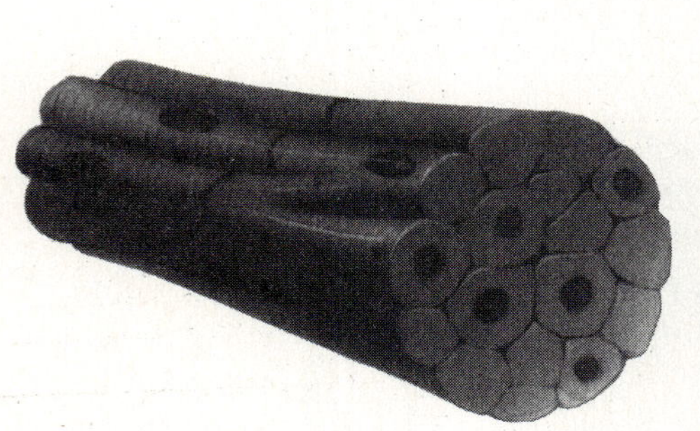

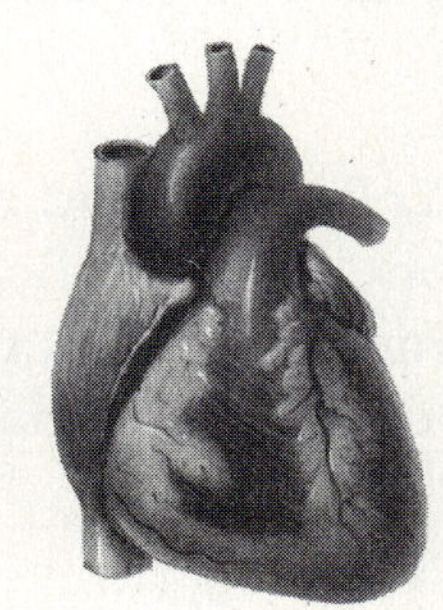

There are muscles that are not under your control. Can you think of any? It is not possible for you to make your heart beat faster just by thinking about it. If you exercise, your heart rate will increase, but normally you can't make it speed up or slow down. **Cardiac muscle**, located in the heart, is one type of muscle that is involuntary. That is, you have no direct control over it.

Smooth muscle that lines the digestive system is another kind of involuntary muscle. Under normal circumstances you cannot speed food through your digestive system, nor can you slow its progress just by thinking about it. Smooth muscle operates even though you don't think about it.

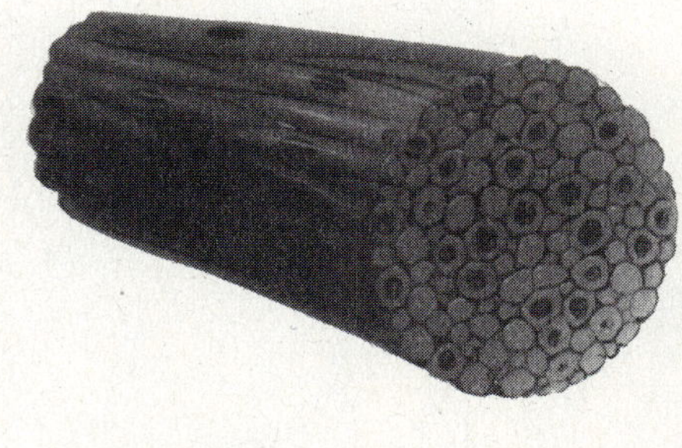

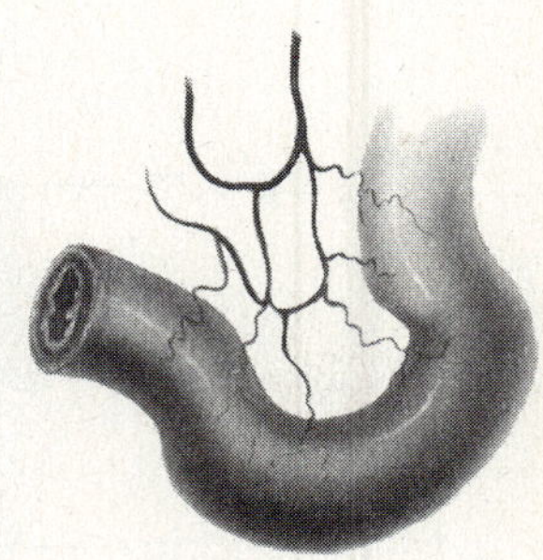

Illustrations not to scale

Minds On! What seems to be the difference between voluntary muscles and involuntary muscles? Why is it good that smooth and cardiac muscles are involuntary? What might happen if they were not? ●

Health Link

How Exercise Improves Overall Health

With exercise you can develop stronger muscles, more energy, and greater endurance. But did you know that there are many other advantages? For instance, exercise strengthens the heart muscle. It also helps strengthen bones and improve digestion. Why might that be so?

Research any type of exercise. Explain the advantages and disadvantages of the particular exercise. Indicate any equipment, precautions, or preliminary exercises that might be necessary. For instance, many sports require warm-up exercises to help avoid injury.

When you've finished your research, prepare a poster, collage, slide show, or other presentation along with your group to explain the exercise you have researched.

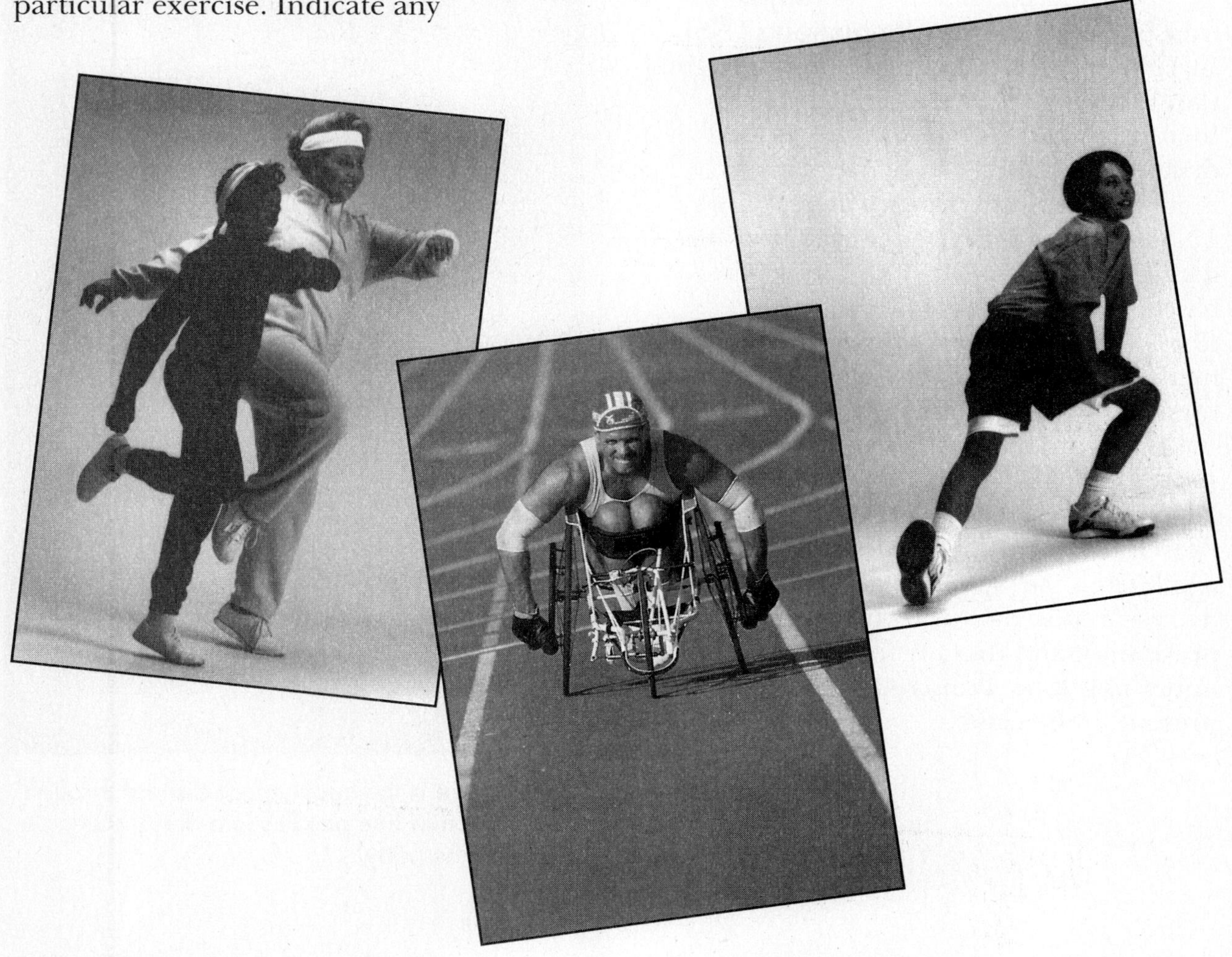

Sports Medicine— Helping Before It Hurts

While exercise is great for your body, if you don't prepare for it properly, or if you overdo it, you may experience an injury. A relatively new field of medicine has developed from a more complete understanding of how the human body works. Sports medicine tries to help people first by preventing injuries. Sports medicine specialists also develop new therapies and conduct studies to understand how to employ old therapies in ways that will speed a person's recovery.

One example of a new therapy is transcutaneous nerve stimulation (TNS). In TNS nerves are stimulated through the skin to relieve muscle pain. Before this therapy, doctors sometimes prescribed drugs that could be addictive, such as morphine and other pain killers. With TNS, some pain can be relieved without dangerous drugs. The TNS therapy is based on acupuncture, which was developed by Chinese doctors. The "maps" made by the acupuncture specialists are used by doctors who wish to use TNS.

There are many other examples of how sports medicine has made everyone more aware of precautions to take during exercise. Make a list of injuries that may result from participation in sports. Then do research to discover what the current prevention and therapy for that type of injury might be. Prepare a short report to present to the class.

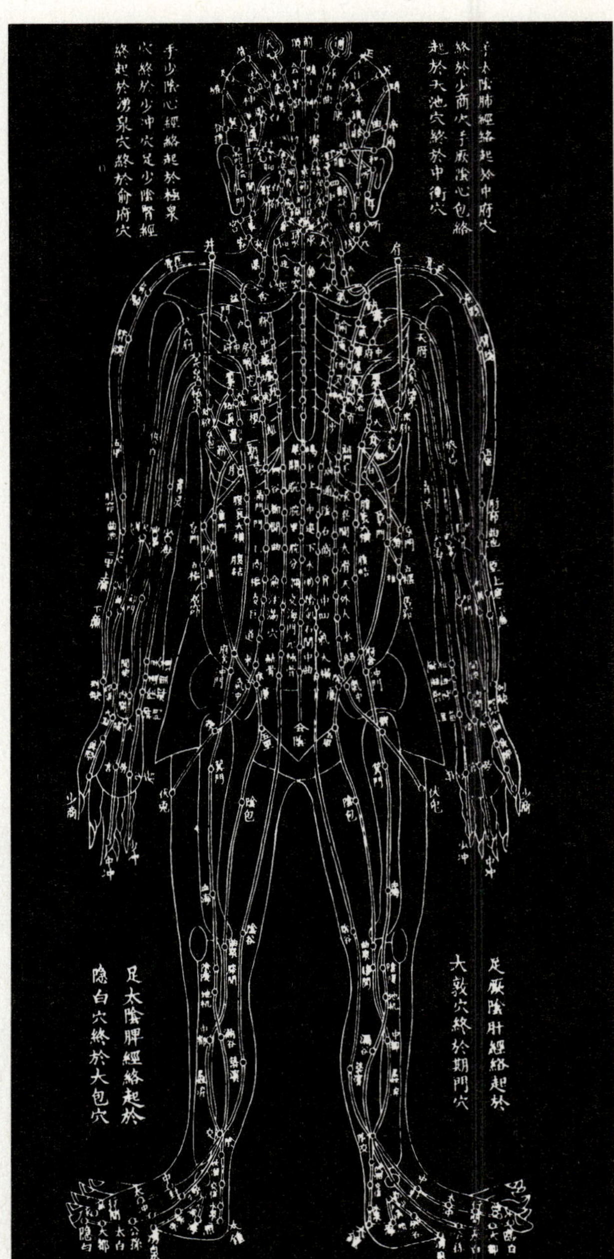

Acupuncture is the treatment of disease through the insertion of fine needles into designated points on the body.

Sum It Up

The skeletal and muscular systems work together and rely on other body systems. Processes within the cells convert the fuel into energy that can be used for movement. The skeletal system also provides you with support, and, in the case of the skull and ribs, with protection. In addition, tissue in bones produces components of the blood such as red blood cells, platelets, and many white blood cells.

Muscles help you move your bones, but they also are vital to a variety of processes. Muscles move food through the digestive system. One of the most important muscles, the heart, pumps blood throughout the body. Without the muscular system, there could be no digestion, no circulation, and no body movement.

Using Vocabulary

Use the vocabulary words to create a crossword puzzle. Make up proper definitions for each word. Give the puzzle to a classmate to solve.

cardiac muscle
compact bone
red marrow
respiration
skeletal muscle
smooth muscle
spongy bone

Critical Thinking

1. How would motion be different for you if you had no bones?

2. If your elbows and knees were ball-and-socket joints, how would you be able to move differently than you can now?

3. Why are bone-marrow transplants sometimes used to treat blood diseases?

4. How can you tell there are more muscles than bones in your arm?

5. Explain how the muscular and digestive systems are important to one another.

Maintaining Balance

Within your community, there are devices to help maintain a stable environment. For example, stop lights, street signs, and the rules of the road help keep the traffic flow stable. Different body systems—including circulatory, respiratory, and excretory, among others—work together to keep your internal environment stable. To perform your daily activities, your body takes in oxygen, food, and water and delivers them to cells to produce energy. It also removes carbon dioxide and other wastes. These systems let your body respond to your needs and decisions throughout the day.

Minds On! You're aware that many body systems resemble systems in your community. List some ways that body systems are like community systems. Compare the way that community systems are regulated with the way body systems are regulated. ●

The circulation or flow of traffic on I-35 near Shawnee, Kansas

Breathe In, Breathe Out

How does your body adjust when you ask it to do work? This activity will give you a clue.

What You Need

stopwatch or watch with second hand
water
container (for water)
stirring rod
graduated cylinder
2 droppers
phenolphthalein solution
dilute sodium hydroxide solution
plastic straw
safety goggles

See the ***Safety Tips*** in steps 4 and 6.

What To Do

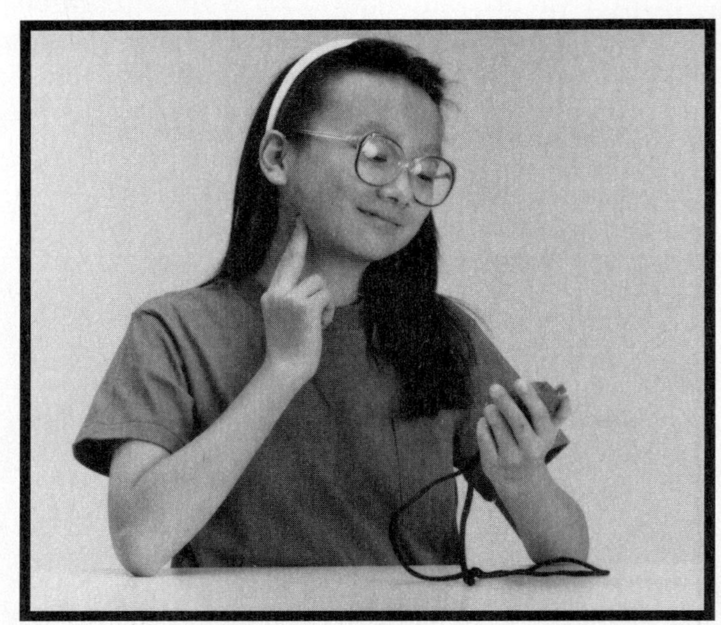

1 Following the instructions your teacher will give you, take your pulse. Record this information.

2 Measure 100 mL of water into a container. Add 4–5 drops of phenolphthalein solution.

3 Add sodium hydroxide solution drop by drop to the container. Stir gently after each drop. Continue to add drops until you notice a change that remains after one minute. What happened to the solution?

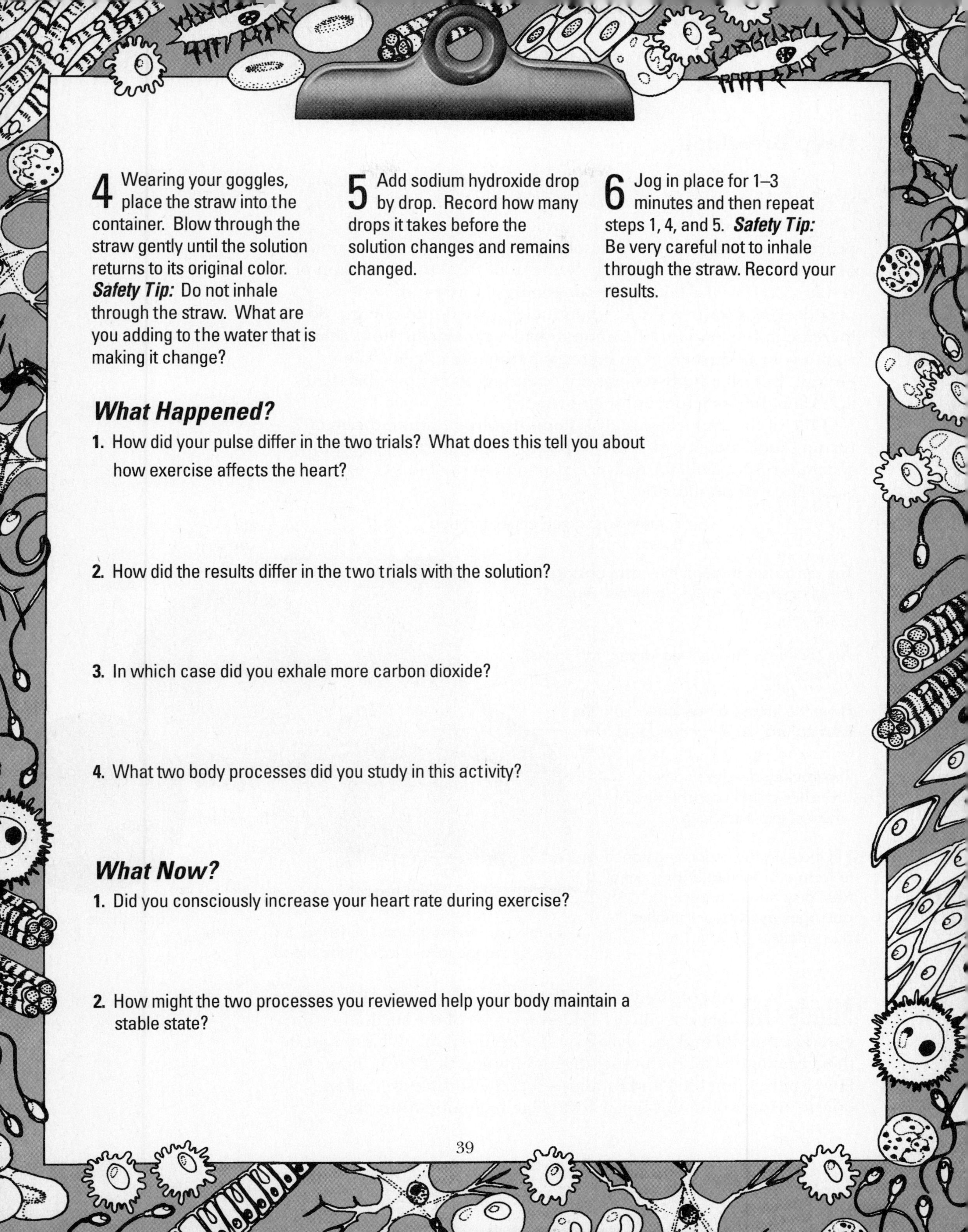

4 Wearing your goggles, place the straw into the container. Blow through the straw gently until the solution returns to its original color. ***Safety Tip:*** Do not inhale through the straw. What are you adding to the water that is making it change?

5 Add sodium hydroxide drop by drop. Record how many drops it takes before the solution changes and remains changed.

6 Jog in place for 1–3 minutes and then repeat steps 1, 4, and 5. ***Safety Tip:*** Be very careful not to inhale through the straw. Record your results.

What Happened?

1. How did your pulse differ in the two trials? What does this tell you about how exercise affects the heart?

2. How did the results differ in the two trials with the solution?

3. In which case did you exhale more carbon dioxide?

4. What two body processes did you study in this activity?

What Now?

1. Did you consciously increase your heart rate during exercise?

2. How might the two processes you reviewed help your body maintain a stable state?

Deep Breathing

In the Explore Activity you saw that your body reacts to changes in your activity. In order for your body to keep up with the energy demands placed on it when you moved faster, some of the body processes also accelerated. You observed an increase in the amount of carbon dioxide you exhaled. Remember from the discussion of respiration how the body generates energy. Carbon dioxide is produced as a waste product when fuel is turned into energy. So an increase in the amount of carbon dioxide you exhale shows that within your body there is an increase in the rate of generating energy. But what body systems are operating to remove this waste and keep the body functioning normally?

One of the important body systems involved in the process of turning fuel into energy is the respiratory system. The respiratory system is responsible for delivering oxygen to the blood and removing carbon dioxide.

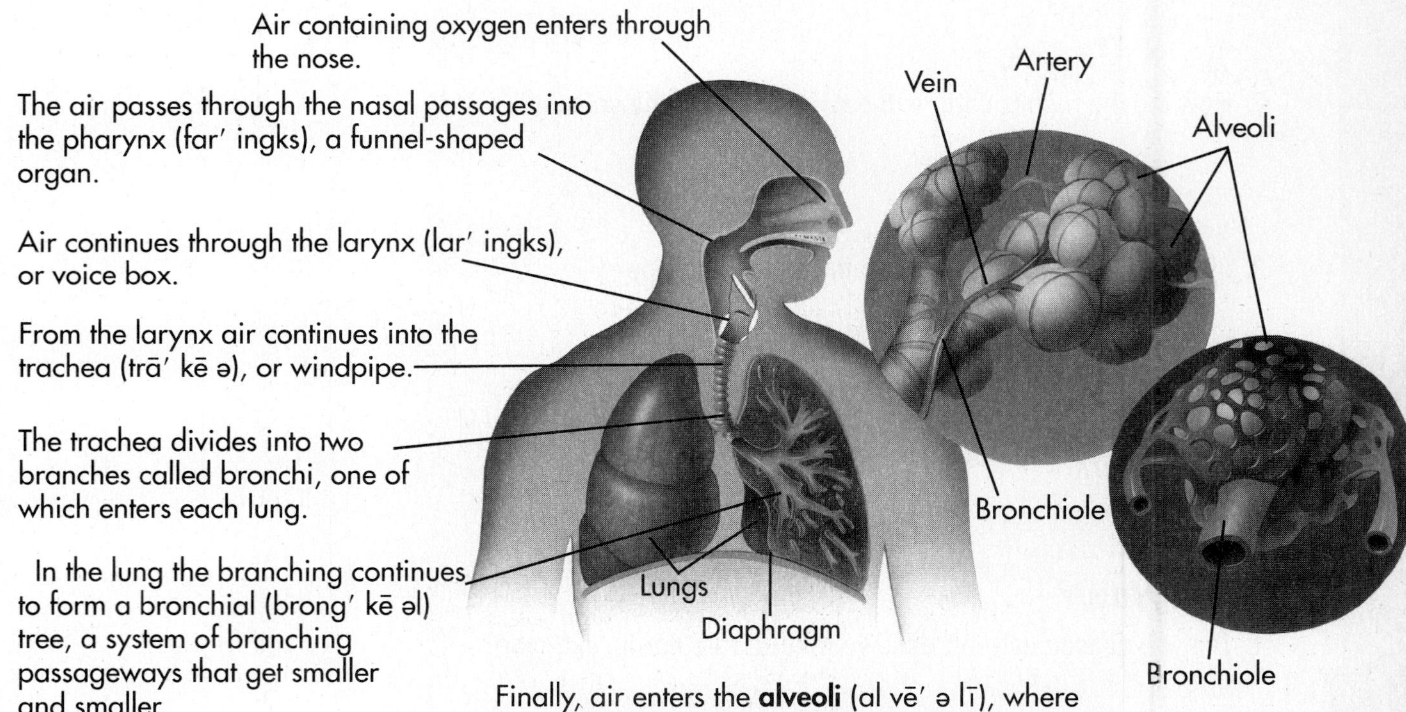

Minds On! Alveoli are very small portions of the lung. Remember that the villi are folded sections of the small intestine. How are the villi and the alveoli similar in function? What might be the advantage of all the branching and folding that form them? How might branching and folding affect the surface area of an object? How would this be an advantage to an organism? ●

The Heart of the Matter

In order for blood to get rid of carbon dioxide and collect oxygen, it must reach the lungs. This is the job of the circulatory system. But how does blood move? The heart, the main organ of the circulatory system, pumps the blood through the body.

The heart, as you remember, is composed mainly of cardiac muscle, which contracts at different times under the control of the nervous system. The heart itself is the core of the pumping process, so you need to examine how the blood moves through the heart in order to understand the pumping action.

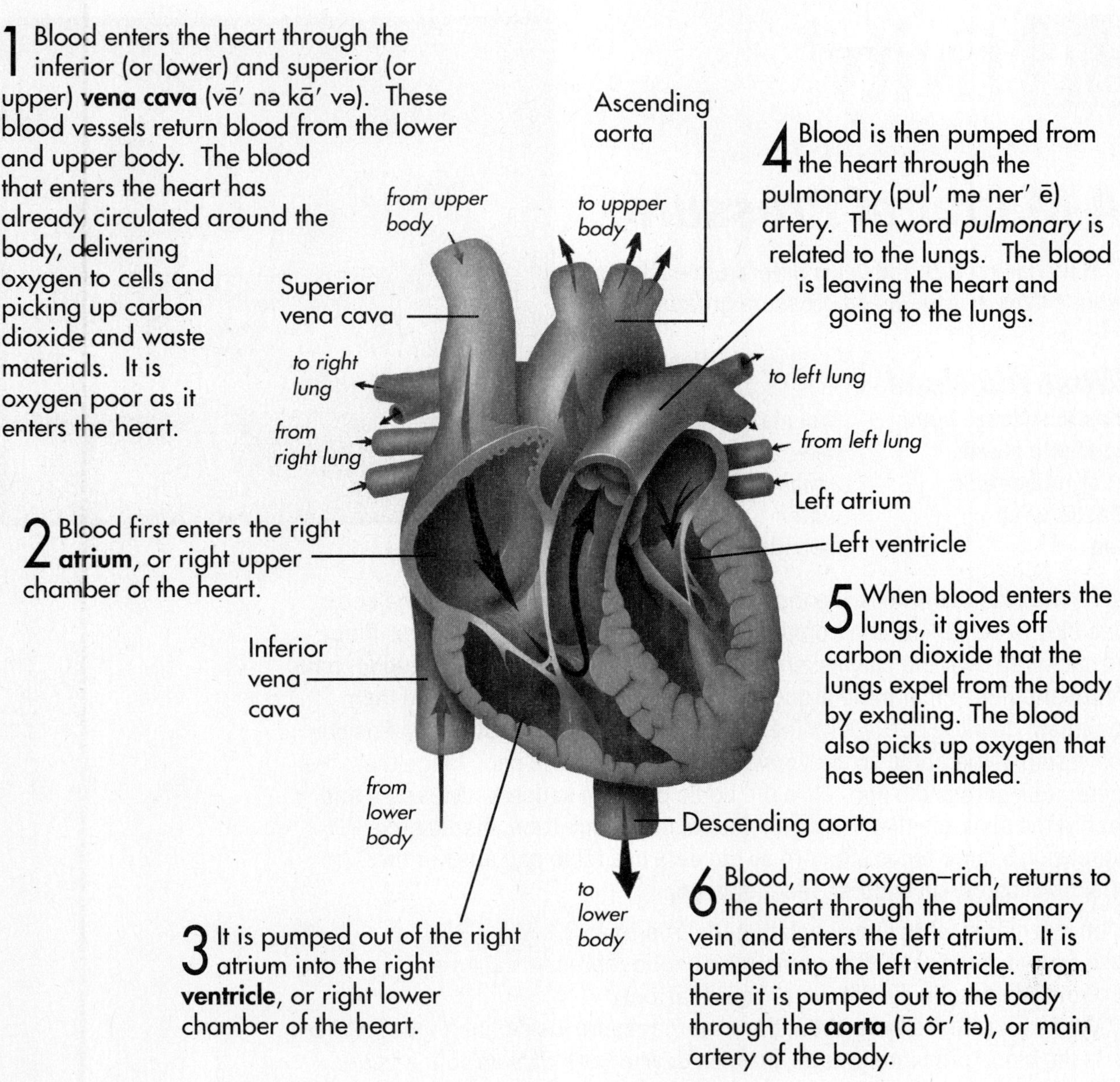

While the heart is the main organ of the circulatory system, there is an entire network of vessels to transport the blood from point to point throughout the body. The vessels through which blood is pumped away from the heart are arteries. Veins are the vessels through which blood returns to the heart. Extremely small and narrow blood vessels through which the exchange of all gases, nutrients, and wastes occurs are called **capillaries** (kap′ ə ler′ ēz).

Do the Try This Activity to explore some of the differences between blood vessels.

TRY THIS Activity!

A Matter of Pressure

How do arteries and veins differ from each other? Where is the blood pressure greater?

What You Need

plastic squeeze bottle
bendable plastic or rubber tube
plastic wrap
pan
firm plastic tube
clay
2 rubber bands
water
meterstick

Fill the plastic bottle to the top with water. Place the 2 tubes in the bottle. Use clay to seal the bottle completely and to hold the tubes in place. Place plastic wrap over the clay assembly and securely attach it with a rubber band. Wrap another rubber band around both tubes near the ends to hold them together. Hold the bottle with the tubes parallel to the pan surface. Position the mouth of the bottle so that it overhangs the lip of the pan. Place the meterstick across the pan. Give the bottle one firm squeeze. Measure and record the distance that the water squirts from each tube. Repeat the squeezing 5 more times. Record all the data in a table and answer the questions below on another sheet of paper.

1. In this model what does each material represent?
2. If arteries are firmer than veins, which tube represents arteries?
3. From which tube did the water squirt farther?
4. What does this tell you about the water pressure inside each tube? Where is the blood pressure greater, in veins or arteries? Explain your answer.

Blood Will Tell

You've had a chance to examine the main parts of the circulatory system—the heart and the blood vessels. These parts function together to transport nutrients, oxygen, and chemicals to the cells of the body and to remove wastes from the cells. They also aid in protecting the body from disease. All of these functions are possible because of the fluid that is pumped through the system. You probably know what blood looks like, and you've read a little about its functions. But what is it made of?

Components of Blood

Platelets—important in blood clotting and a component of scabs, under which new skin cells can grow to close a wound.

Red blood cells

Red blood cells—carry oxygen and nutrients to cells and remove waste from cells. Red blood cells make blood look red. **Hemoglobin** (hē′ mə glō′ bin) is a complex protein that includes four iron atoms. Oxygen molecules attach to the iron molecules. There are about 280 million hemoglobin molecules per red blood cell.

White blood cells—responsible for fighting off infections in a variety of different ways.

Plasma—the nonliving, straw-colored, liquid part of the blood in which the living cells are suspended.

Health Link

You're Just My Type

As you may know, people undergoing surgery or recovering from wounds often need blood transfusions. However, a person can't donate or receive blood from just anyone. Receiving the wrong type of blood can cause clotting.

All humans have blood that occurs in one of four types. Blood types are A, B, O, and AB. Any two healthy people with the same blood type can safely donate blood to one another. Research how blood typing is done and the other factors that are looked for during typing. Report your results to the class.

Founder of the Blood Banks

Charles Drew researched and understood blood and its functions before World War II. During the war he set up blood banks in England. He also encouraged doctors to give transfusions of plasma rather than whole blood. Plasma does not need to be typed, and it can be donated and received by anyone without negative effects. His blood banks and plasma transfusion technique are credited with saving millions of lives, both during the war and in the years afterward. Tragically, Dr. Drew died for lack of a blood transfusion after an accident.

Research the life of Dr. Charles Drew or another pioneer in the study of blood and circulation such as William Harvey. Make a poster or series of "slides" and present your findings to the class.

Dr. Charles Drew

In With the Good, Out With the Bad

You know that blood removes wastes from cells. But then what does it do with them? In the lungs carbon dioxide, a waste product of respiration, is exchanged for oxygen, which is carried to body cells. But what happens to the other wastes from cell processes? How are they removed from the body?

There are several different kinds of waste that are removed from the body. One kind of waste is dead and dying blood cells. These are removed and destroyed by the liver.

Some wastes are removed through the skin when you sweat. However, most wastes aren't removed through the skin. They are removed by the kidneys.

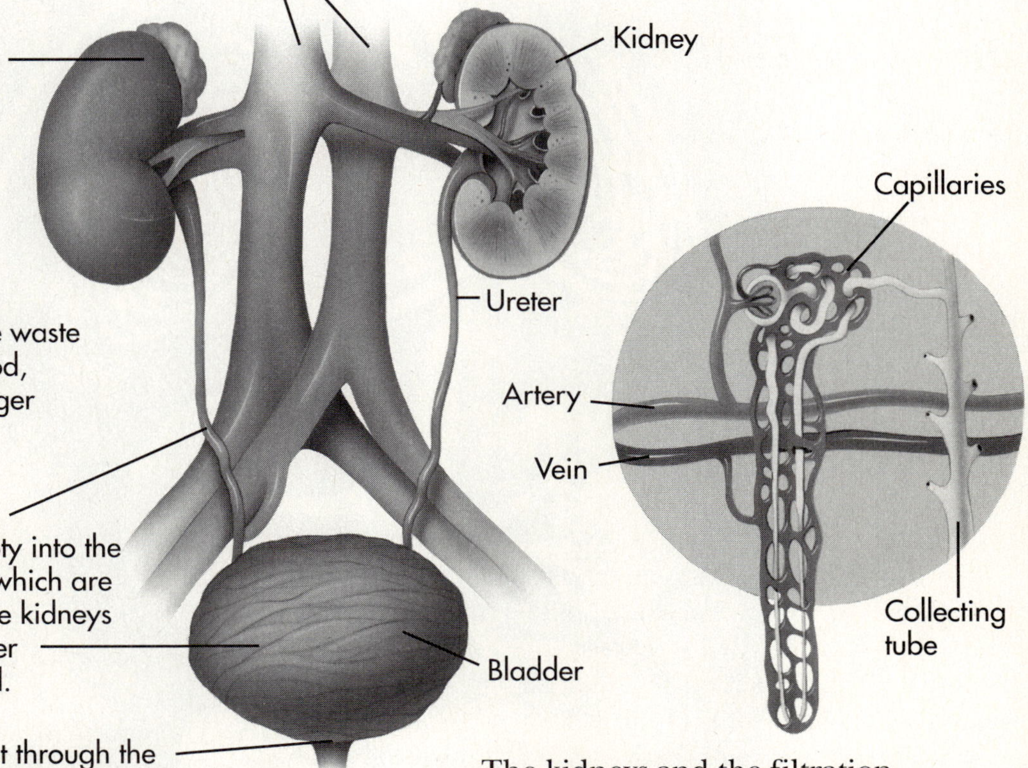

Blood enters the kidneys through the renal (rē′ nəl) arteries.

Blood continues through smaller and smaller arteries and finally enters a small artery in a bulb. Here the waste is filtered out of the blood plasma and enters the urinary tubule.

These tubules collect urine (yur′ in), or the waste filtered from the blood, and channel it to larger collecting ducts.

All of the collecting ducts eventually empty into the ureters (yu̇ rē′ ters), which are tubes that connect the kidneys to the urinary bladder where urine is stored.

Urine passes out through the urethra (yu̇ rē′ thrə) and is expelled from the body.

Kidney

Ureter

Bladder

Capillaries

Artery

Vein

Collecting tube

The kidneys and the filtration apparatus are very precise, filtering out only the wastes and not large, useful proteins. In addition the kidneys refilter the wastes and return most of the water to the blood. If they did not do so, dehydration would occur. All blood is completely filtered by the kidneys once every 40 minutes.

Around the Body in a Minute

Minds On! You've already learned a great deal about how the human body works. Here's a chance to put it all together.

A blood cell travels through the body in about one minute. On the diagram below, trace the journey of the blood cell as it moves from the small intestine where it has absorbed nutrients through the heart, to the lungs, and to a muscle cell in the arm of a person swimming. Then trace the path back from the muscle cell through the heart to the kidney. Indicate what is happening to the blood cell in the small intestine, heart, lung, muscle cell, and kidney. ●

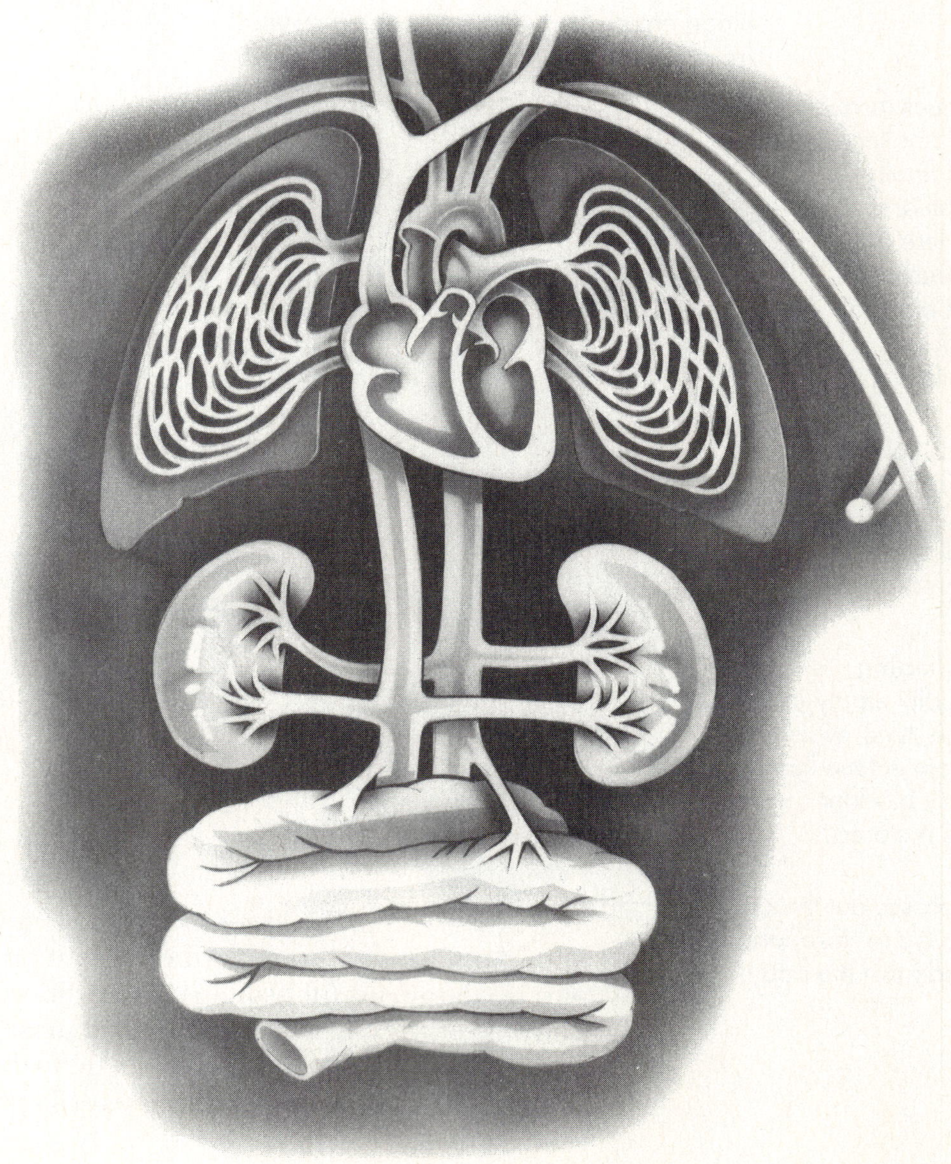

What Can Go Wrong?

Minds On! You can see how diseases and disorders that affect vital body systems would be particularly hazardous to overall health. List some of the effects on health you would expect from diseases that harm the body's ability to obtain and transport oxygen to all body parts. What might be the effects of diseases that impair the body's ability to remove wastes? ●

Health Link

Actions for Health

You can greatly reduce the chances of getting some of the diseases and disorders that affect the regulatory systems of your body. For instance, the risk of coronary artery disease, which can lead to a heart attack, is much less if you do not smoke. You may be aware that smoking can cause lung cancer. Smoking may also cause or aggravate conditions like emphysema, a disease that destroys the alveoli in the lungs. Smoking may contribute to high blood pressure, which can lead to problems in all the organs of the body.

Likewise, a poor diet can contribute to problems with the heart and blood vessels. Arteriosclerosis (är tîr′ ē ō sklə rō′ sis), or hardening of the arteries, can result from a diet high in fat and cholesterol. Arteriosclerosis can lead to blockages of the arteries in the heart and elsewhere in the body. Blood clots and heart attacks are possible results.

Kidneys can be severely affected by high blood pressure. Factors that lead to high blood pressure or increase its effects may also cause kidney damage.

In a group, research one of the diseases or disorders mentioned above or another disease or disorder related to the systems you have studied in this lesson. Prepare a report in which you describe the causes of the disease, the symptoms, the effects on the body, and steps that you can take to prevent or lessen the chances of getting the disease or disorder. Present your information to the class in the form of an interview or newscast.

Buildup of fatty material in blood vessels can lead to blockages that result in poor circulation, blood clots, and sometimes heart attacks.

Hunting Down the Hairy Cell

Dr. Bouroncle's efforts have brought us one step closer to identifying and eliminating one form of cancer.

You can reduce your chances of coming down with some diseases and disorders. However, there are some diseases and disorders that you cannot control. For these diseases we rely upon the work of doctors and other health researchers to develop treatments and therapies.

Dr. Bertha Bouroncle (brôn′ clā) is one of a team of researchers who, in 1958, identified a type of blood disease called hairy-cell leukemia. It is a disease that affects one type of white blood cell. As a result of the disease, the affected cells grow long hairlike projections. The person who has the disease suffers from fatigue, infections, bleeding, and develops enlargement of an organ called the spleen. At the time it was discovered, the only treatment was surgery to remove the enlarged spleen. In about half of the cases, the patients improved. This success rate was not good enough, since doctors are working towards 100 percent success.

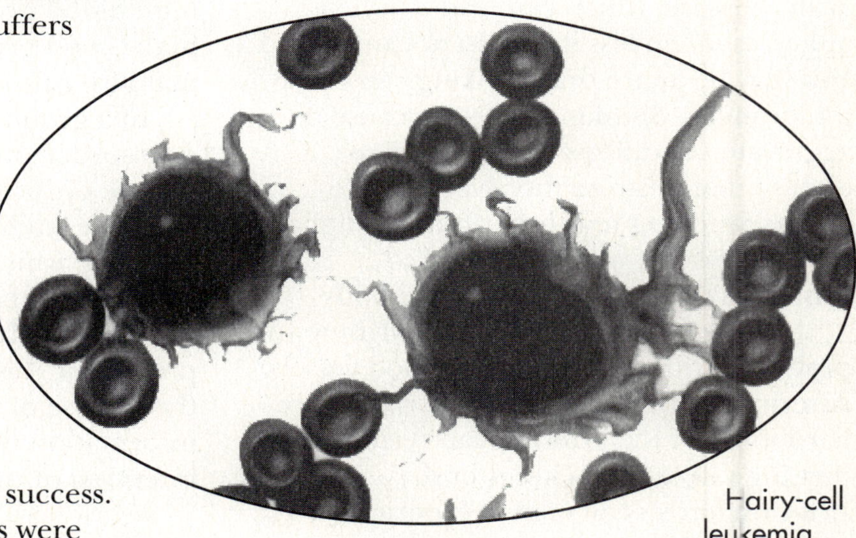

Hairy-cell leukemia

Many different treatments were developed by researchers over time. They were tested in patients with hairy-cell leukemia by many investigators, including Dr. Bouroncle.

Recently two forms of chemotherapy, or treatment using drugs, have increased the success rate of treatment to 85–90 percent. In fact, hairy-cell leukemia may soon be considered curable.

The efforts of scientists and doctors like Dr. Bouroncle have led to treatments for a great many diseases and disorders. Technology, such as development of new drugs, must be used together with understanding of chemistry and of the human body to develop new treatments for all kinds of diseases.

Sum It Up

The circulatory, respiratory, and excretory systems are responsible for regulating and maintaining a stable internal environment in which body functions can continue. The circulatory system connects all the other systems in the body. It provides nutrients that it picks up from the digestive system and oxygen that it picks up from the respiratory system. Blood cells deliver these materials to cells in the body and pick up wastes that are removed from the body through the excretory system. Through the interactions of these three systems, your body continues to function day after day.

Using Vocabulary

Draw and label a diagram that shows the location of each of the vocabulary words within the human body. (Hint: you'll need to draw a cell to show the location of one of them.)

alveoli	capillaries	vena cava
aorta	hemoglobin	ventricle
atrium	platelets	

Critical Thinking

1. Explain how the circulatory system interacts with all the other body systems.

2. Why might creating the greatest possible surface area be important in organs like the lungs?

3. Why might high blood pressure be destructive to body tissues?

4. What might be the result of not having enough platelets in your blood?

5. Carbon monoxide, which is found in cigarette smoke, bonds tightly to hemoglobin. How would this impair the functioning of your body?

Securing the Borders

Skin, hair, and nails—your body's first line of defense—make up your body's armor. This armor is a covering designed to keep foreign particles out.

Foreign particles are any particles including bacteria, viruses, fungi, dust, and other things that enter the body from outside and provoke a response from the body. If foreign particles get past your first line of defense, there is a second, internal line of defense that protects your body from harm. The second line of defense takes several forms and is the result of several systems interacting to serve one purpose. It may not seem that there's much to prevent damage to delicate body systems, but you might be surprised.

Protective armor

Minds On! Think about protection and devices designed to protect humans. Make a chart with two columns. In one column list all of the devices that are designed to protect humans in some way. These might include anything from signs to seat belts. In the second column list how the device works to provide protection. ●

Now that you've got an idea of how some protective devices work, do the activity that follows to examine the first line of body defenses.

One Bad Apple

You may have heard that one bad apple spoils the whole bunch. Well, here's your chance to see if the old saying is true. You'll also be able to find out how you can stop the influence of one bad apple.

What You Need

5 fresh apples
decaying apple
soap
water
paper towels
plastic knife
rubbing alcohol
labels
5 self-sealing plastic bags
pen
rubber gloves

What To Do

1 Label your bags with numbers 1 through 5. Then, put on the rubber gloves.

2 Rub the decaying apple all over the outside of each fresh apple. Throw away the decaying apple. Wash your hands while you're wearing the rubber gloves.

3 Place a fresh apple in the bag labeled ***1***.

4 Wash one apple with soap and water. Place it in the bag labeled ***2***. Drop an apple on the floor so that it is bruised. Place this apple in the bag labeled ***3***. Use the plastic knife to punch several slits into an apple. Place this apple in the bag labeled ***4***. Rub the outside of the last apple with alcohol. Place this apple in the bag labeled ***5***. Now take off the rubber gloves and wash your hands.

5 Seal all of the bags.

6 Store the apples in a dark place. Predict what will happen to each apple. Record your predictions.

7 Make a chart to record your observations of the five apples over the next seven days. Each day examine each of your apples carefully. Do not open the sealed bags. Note any changes in color, appearance, or texture. When you are finished, return the apples to their storage place.

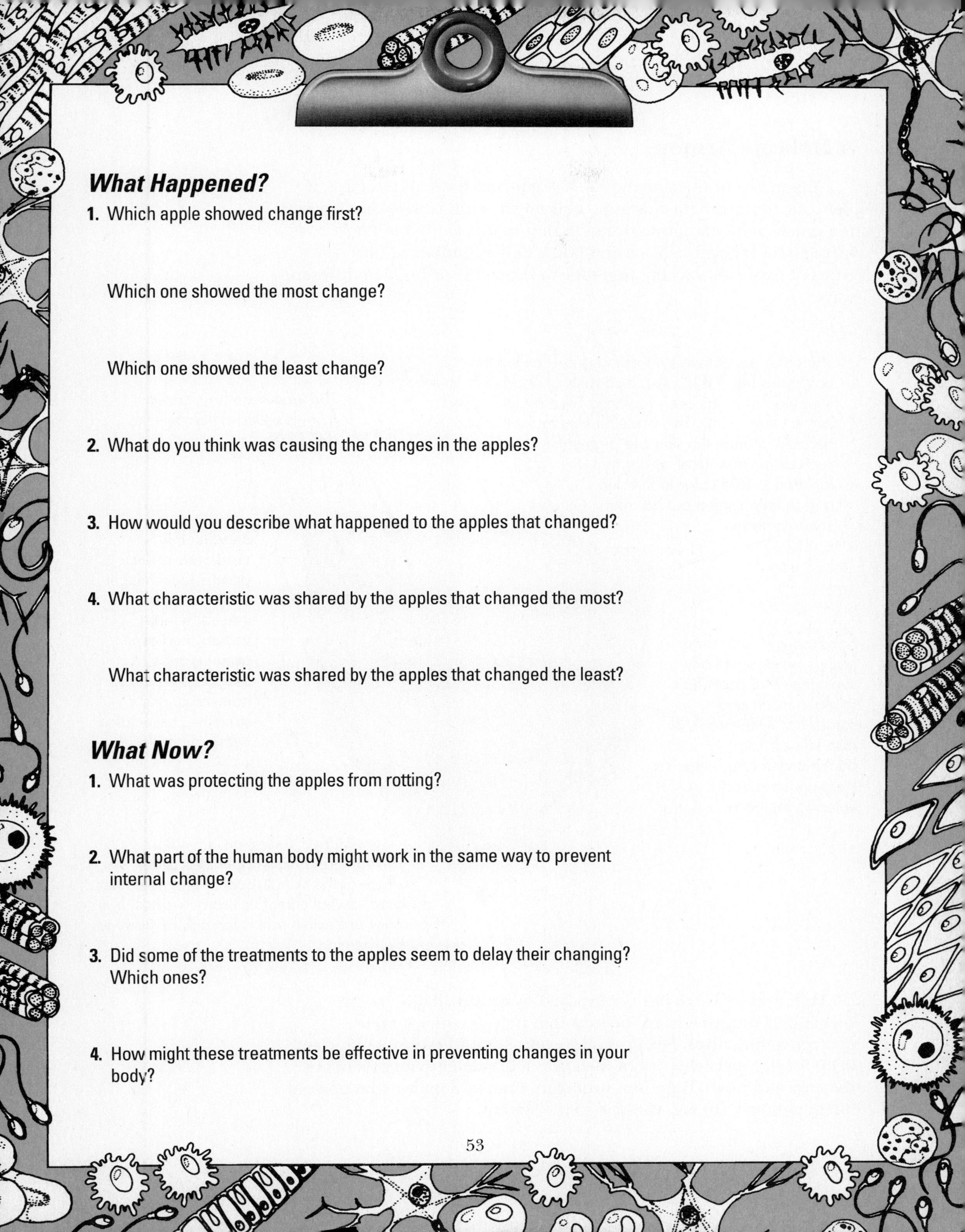

What Happened?

1. Which apple showed change first?

   Which one showed the most change?

   Which one showed the least change?

2. What do you think was causing the changes in the apples?

3. How would you describe what happened to the apples that changed?

4. What characteristic was shared by the apples that changed the most?

   What characteristic was shared by the apples that changed the least?

What Now?

1. What was protecting the apples from rotting?

2. What part of the human body might work in the same way to prevent internal change?

3. Did some of the treatments to the apples seem to delay their changing? Which ones?

4. How might these treatments be effective in preventing changes in your body?

A Delicate Armor

As you saw in the activity, many organisms have an effective, if delicate, defensive mechanism. This mechanism is the skin. The skin is a much more elaborate organ than it might seem on first inspection. It consists of several layers and a number of different types of tissue all working together to protect your body in different ways.

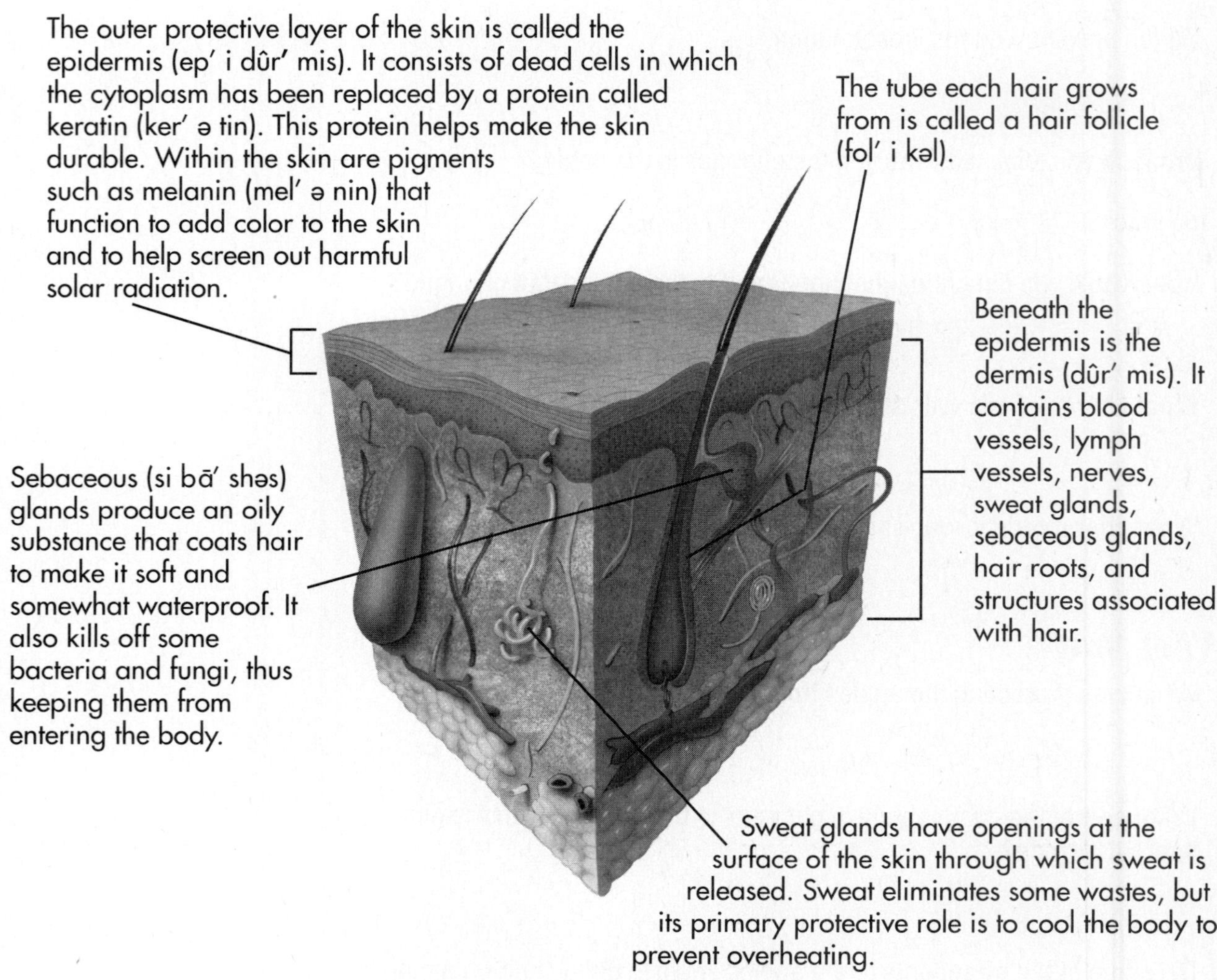

Hair and nails are both composed of keratin. They are specialized outgrowths of the skin that help provide similar protective functions. For instance, eyelashes and eyebrows help to keep foreign particles out of the eyes. The same is true of hairs in the ears and nose. Together, your skin, hair, and nails make up your **integumentary** (in teg′ yə men′ tə rē) **system.**

TRY THIS **Activity!**

How Can Fingerprints Be Helpful?

You've learned something about the skin, and you probably already know that people have fingerprints. Each person's fingerprints uniquely identify that person. But other than identification, what else could fingerprints be useful for?

What You Need

hand lens	**doorknob**	**cloth work gloves**
pencil	**coins or paper clips**	

Use the hand lens to examine your fingertips. On another sheet of paper, draw a picture of what you see. Put on the work gloves and attempt to pick up the coins or paper clips. Pick up the pencil and try to write. Now try to turn the doorknob.

1. Were you able to hold the pencil and write?

2. Were you able to pick up the coins? Turn the doorknob?

3. How did the gloves affect your ability to perform each action?

4. Develop a hypothesis that suggests one function for the ridges and folds on your fingers.

As you saw, the ridges on skin do serve a useful function. These fingerprints can be divided into three main groups depending on their pattern. Examine the pictures of the types below and see if you can identify what pattern your fingerprints have.

Arches

Loops

Whorls

The Lymphatic System

One of the important parts of the immune system and of several other systems is the **lymphatic** (lim fat′ ik) **system.** The lymphatic system is similar to the circulatory system in function. It transports nutrients too large to enter the bloodstream (such as some large fat molecules) and removes waste.

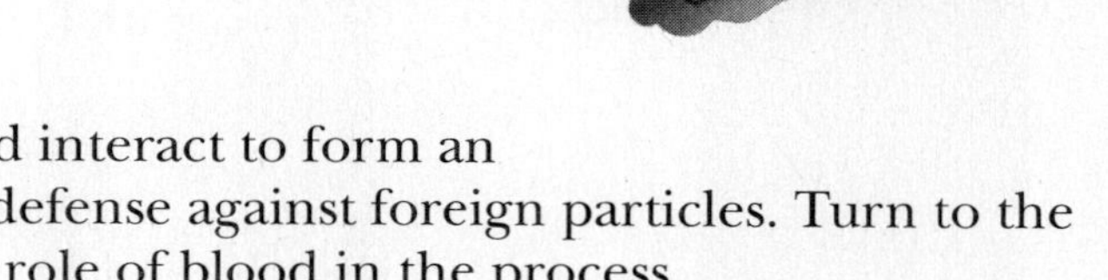

Lymph and blood interact to form an incredibly complex defense against foreign particles. Turn to the next page to see the role of blood in the process.

Mobilizing the Defenses

What happens if foreign particles such as bacteria, viruses, and fungi get past the skin? Mucous (mū′ kəs) membranes lining the nose and respiratory tract produce mucus, which is a sticky substance. Foreign particles that come into contact with the mucus stick to it. They are removed as the mucus is removed. In the trachea tiny hairs called cilia move the mucus toward the pharynx where it is either swallowed or spit out. This can often be accompanied by coughing, which also helps to remove mucus.

If particles manage to get past this second line of defense, the body has an elaborate system to fight off foreign particles such as bacteria, fungi, and viruses. The **immune** (i mūn′) **system** functions to fight off invaders.

The immune system is a complex system of cells and chemicals that helps the body fight off infection.

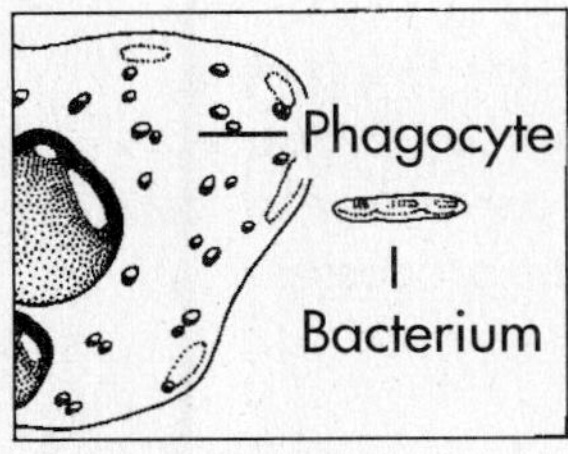

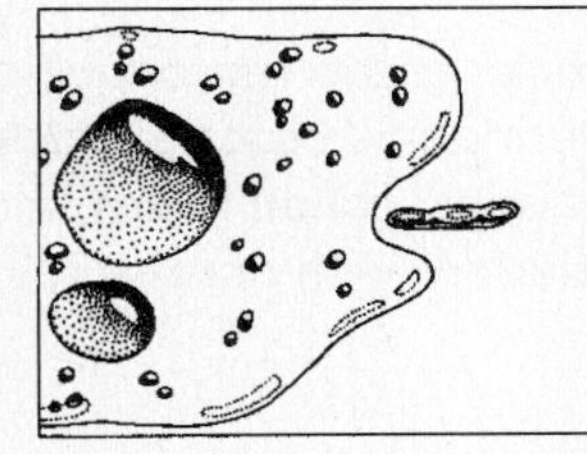
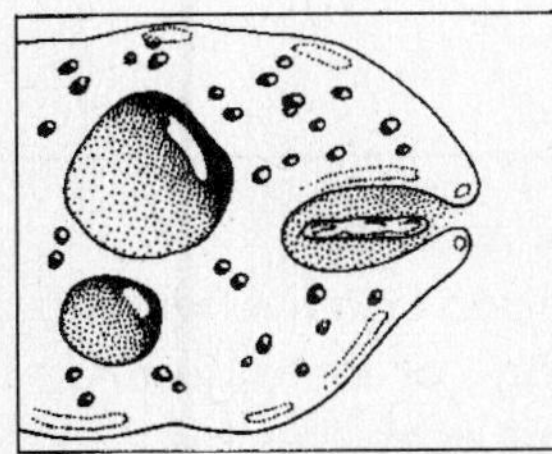
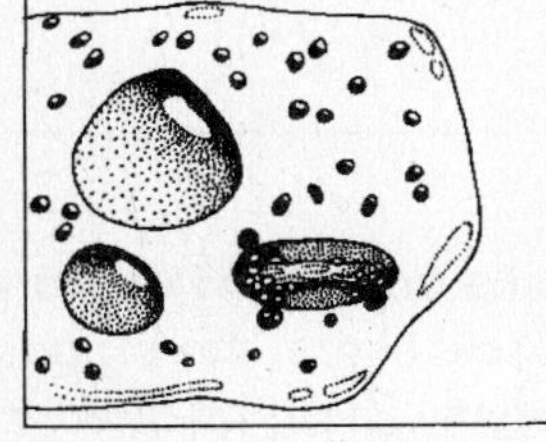

Phagocytes (fag′ ə sīts), a type of white blood cell, swallow up foreign particles in much the same way an amoeba surrounds and engulfs its prey. The liver has a large number of phagocytes that remain in place. As the blood is filtered through the liver, the phagocytes remove almost all of the foreign particles.

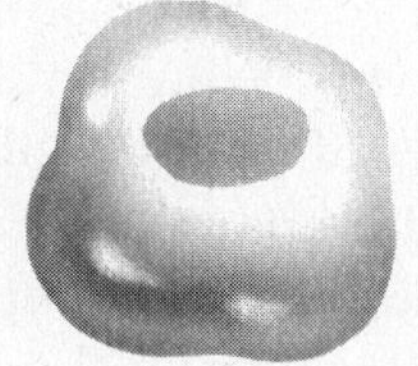

B cell

B cells, another kind of white blood cell, mark foreign particles with proteins called antibodies. These proteins cause other complex proteins in the blood, called complement, to move in and destroy the foreign particles.

Killer T cell

T cells come in many varieties. The action of some of these is not clearly understood. However, killer T cells defend the body by destroying body cells that are infected with foreign particles. Killer T cells are often responsible for tissue rejection in transplant operations.

Helper T cell

Helper T cells help B cells fight off invaders.

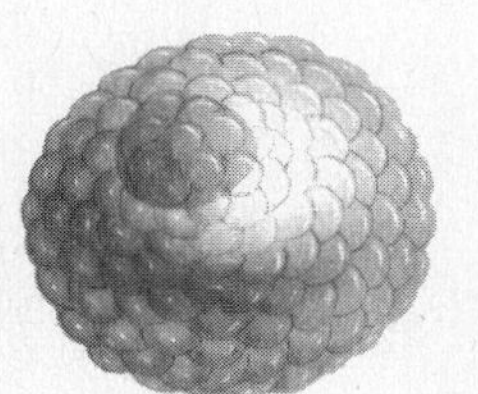

T suppressor cell

T suppressor cells alert the body that the foreign particles have been destroyed and the defenses can relax.

Strategies of Invaders

Minds On! Now you've learned something about how the body protects you. What do you think it's protecting you from? On another sheet of paper, make a chart with three columns. In the first column of the chart, write the names of three different kinds of foreign particles. In the second column of the chart, write how you think they work to make the body sick. In the third column of the chart, write the name of a disease that may be caused by this type of foreign particle. ●

How do foreign particles and organisms harm the human body? There are several ways in which damage can be done.

Very small particles, such as viruses, can inject their genetic material into living cells and begin to reproduce. As viruses reproduce they fill the cell until it bursts, killing the cell and spilling virus particles into the blood and lymph. Bacteria and other foreign particles can also cause cell destruction.

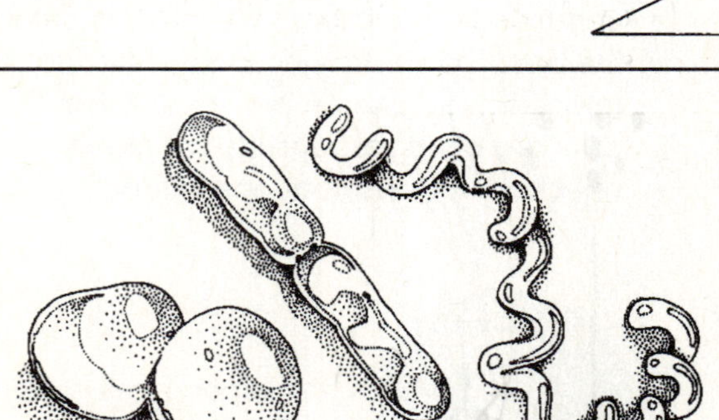

Bacteria can reproduce outside living cells. During this process, bacteria can produce poisons, called toxins, that can either kill cells or interfere with normal cellular processes.

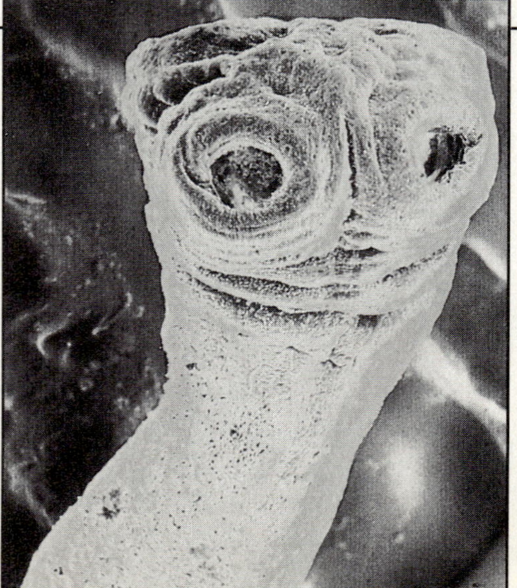

Some organisms are larger and can live within muscle tissue, the bloodstream, or the digestive system. These invaders create problems by giving off foreign wastes into the body and by consuming nutrients meant for body cells.

Tapeworms are just one kind of parasite that can live within the human digestive system.

Illustrations not to scale

Language Arts Link

How Does the Immune System Respond to Invaders?

The response of the immune system to foreign particles is quite complex. Sometimes acting out a process can help you understand it better. Use the information below and any other information you can find to write and act out a play in which the immune system responds to a bacterial infection. A couple of members of your group might write the dialogue that will explain the actions. One or two may design costumes or sets. Others can work on casting and staging the play. When you are finished, perform the play for another class to explain the functioning of the immune system.

The following are steps in the functioning of the immune system:

- A foreign particle, such as a virus or bacterium, enters the bloodstream.
- Phagocytes survey the problem and alert helper T cells.
- Helper T cells multiply and alert the B cells.
- B cells produce antibodies that attach to the invading cell.
- Blood proteins recognize and surround the foreign particle that has the antibodies attached.
- The foreign particle is destroyed by the blood proteins.
- T suppressor cells move in and alert the body that the foreign particle has been destroyed.

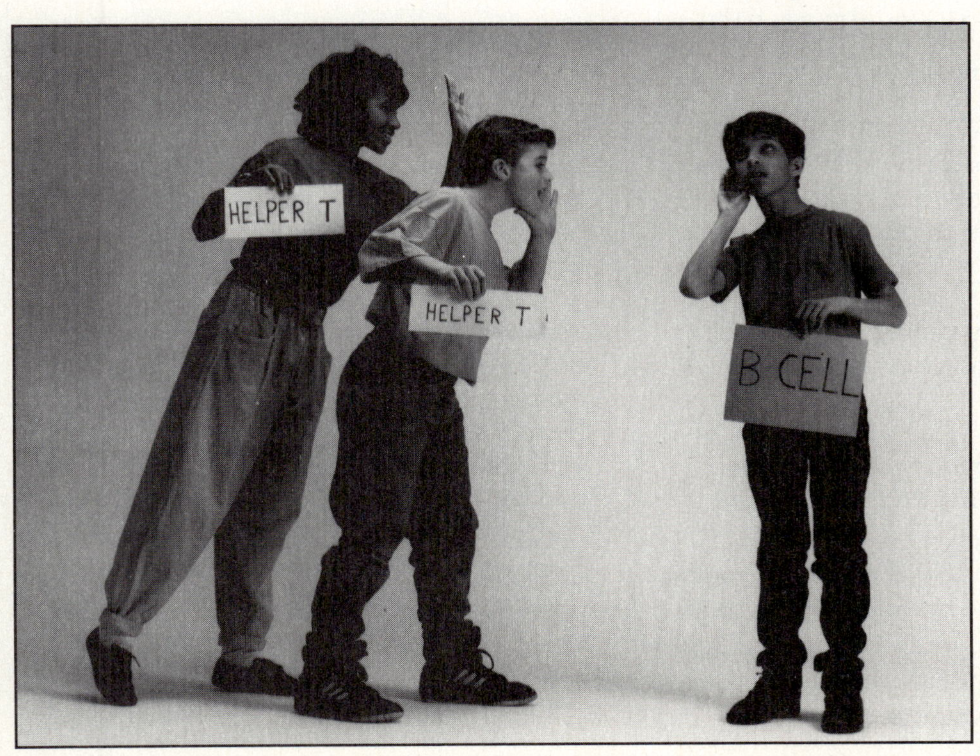

Types of Immunity

It's important to remember that the immune system responds to invaders. It responds more quickly to invaders that it recognizes. In fact, sometimes it reacts so swiftly to a foreign particle it recognizes that you do not come down with the disease. You are immune! **Immunity** is the body's resistance to disease.

Use the chart on this page and the information below to organize your thoughts about how the immune system becomes able to recognize and fight off an invader. When you are through organizing your information in the chart, draw a simple diagram on a sheet of paper to help you understand the relationship of the types of immunity to one another.

Type of Immunity	How It Occurs in the Body
Natural immunity	
Acquired immunity	
Active acquired immunity	
Passive acquired immunity	

- Natural immunity is a type of immunity that results when you have had no known exposure to the foreign particle that causes the disease.
- Acquired immunity is immunity that results from exposure to the disease-causing foreign particle.
- Active acquired immunity can occur when you contract a disease and your immune system responds and destroys the invader. It can also occur when you get a vaccination or shot. The vaccination is made from dead or weakened foreign particles.
- Passive acquired immunity can occur if antibodies were transferred from your mother to you though the placenta before birth. It can also occur if you are injected with antibodies that were produced in another body.

Minds On! Make a list of all the diseases you can think of that you can be vaccinated against. Then list those diseases that you have been vaccinated against. Use reference books to see if you can figure out if these are diseases caused by bacteria, viruses, or fungi. ●

Deficiencies and Drugs—The Hazards of Growing Up

In Lesson 1 you looked at some of the diseases that could result from deficiencies of nutrients in your diet. Research more of these deficiency diseases and try to answer the question, *What would happen if these diseases occurred during childhood?* What might be the effect of such a disease on a person's entire lifetime if it developed early in life? Share your results with the class.

As you are aware, drugs can also have strong effects on development and growth. What effect might exposure to cigarette smoke and alcohol have on a person who is not yet fully developed? Research the effects of both tobacco and alcohol on people. Hypothesize what might occur if these effects were present early in life.

On a separate sheet of paper, use a straightedge to make three columns. In one column list the various drugs you know about or have heard of. In a second column, list the effects of these drugs on the human body. In a third column, list the possible effect of the drug on future generations. For instance, how might a mother's smoking affect the development of the baby within her?

As a class share your information. Imagine that you are members of a government council that is going to make recommendations to the president regarding measures to take to reduce the health risk of nutrient deficiencies and drugs, particularly on young people. Think about what measures you can take at a national, state, and local level. Present your plan to another class, or write a short summary of it and share it with the school.

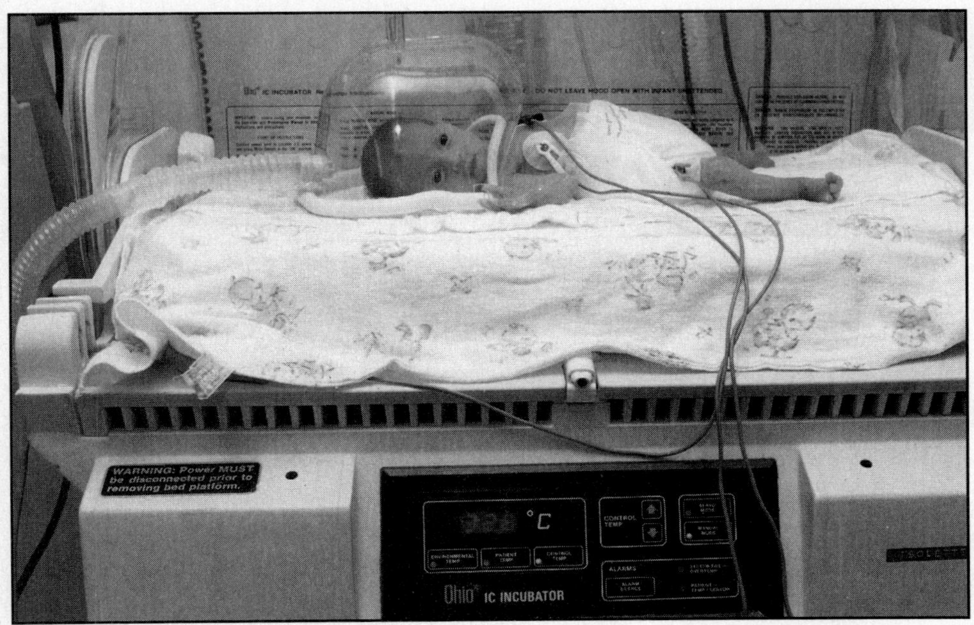

Mothers who smoke, drink, or abuse drugs during pregnancy increase their chances of having babies who suffer from low birth weight, incomplete development, and other problems.

Sum It Up

Each person grows from the union of a cell from the mother and a cell from the father. Through many stages of development, first within the mother and then outside the mother, a person grows and develops. Eventually, through the process of growth and development, the person becomes capable of reproducing. There are a great many changes in the human body as a person approaches reproductive maturity. Reproduction ensures that the human species will continue. A person continues to develop and change through time. Eventually a person will die and thus complete the individual human life cycle.

Using Vocabulary

Use the vocabulary words to write a paragraph that traces the life cycle of a human.

childhood	ovaries	testes
endocrine system	puberty	
infancy	reproductive system	

Critical Thinking

1. If a person did not go through puberty, would he or she be able to reproduce? Why or why not?

2. What might be the advantages of having two parents contribute to offspring rather than just one?

3. Would a person born without a reproductive system be able to survive? Why or why not?

4. Explain how hormones contribute to growth and development.

5. Why might it be a good idea for a pregnant woman to check with her doctor before taking any medication?

Theme T SYSTEMS and INTERACTIONS

The Sensory Web

Within your body is an amazing chemical system. This system gathers facts from both inside and outside your body. Are you hungry? Tired? Chilly or overheated ? Anxious? Cheerful? Whatever state you're in, this system reacts to the chemical messages and tells your body how to react–whether it's by shivering, perspiring, yawning, or hunting for food.

This system even controls whether you are awake or sleeping. It keeps your heart beating and maintains your breathing rate. On command it will activate your muscles to play a guitar, write your name, or dance. It coordinates all the other body systems. For example, if you order your body to take a hike right after lunch, this system reroutes blood from your digestive organs to the muscles of your legs where more oxygen is needed in a hurry. This, the body's master control system, is the nervous system.

Minds On! With several classmates, list the types of information that your brain processes and reacts to during a typical lunch period. Include everything you can think of. When you're finished making your list, group the items by the senses that are involved in each. You may wish to have a grouping into which you put all items that do not fit elsewhere. Share these items with your class and see if other classmates have some ideas about how to classify them. ●

The nervous system coordinates the body and keeps it alive. Continue reading this lesson to learn more about how this complex system works.

How Fast Do You React?

You know that the nervous system senses what is happening around the body and decides what to do about it. Within the nervous system, many different sensations and pieces of information are brought together and studied. Watch your nervous system at work in this activity.

What You Need

coin
electronic stopwatch
metric ruler
colored pencil

What To Do

1 Start the electronic stopwatch. Let the stopwatch run until the seconds indicator reads *5.* Stop the stopwatch. Record the time shown on the watch. Repeat for a total of 10 times. Average the times indicated on the stopwatch.

2 Place a coin on the back of your hand. Tilt your hand so the coin slides off. Attempt to catch the coin with the same hand. Record whether or not you caught it. Do this 10 times with each hand, and record the results of your trials.

3 Have a partner hold a ruler straight up and down so that the bottom end is between your thumb and forefinger. Watch the bottom of the ruler. When your partner drops the ruler, try to catch it. Record the distance the ruler fell before you caught it. (If you didn't catch it, record the entire length of the ruler.) Repeat this step for a total of 10 trials. Average the distance the ruler fell for the 10 trials.

▼

4 Graph the information in steps 1 and 3. Place the trial number along the bottom of the graph and list the seconds passed or the distance fallen in each trial along the side of the graph. On each graph draw a line in colored pencil that indicates the average for all of the trials.

Seconds passed: 5, 6, 7, 8, 9, 10
Number of trials: 1 2 3 4 5 6 7 8 9 10

Step 1

Distance in cm: 3, 6, 9, 12, 15
Number of trials: 1 2 3 4 5 6 7 8 9 10

Step 3

Answer the questions below on another sheet of paper.

What Happened?

1. What systems were functioning together to complete each section of the activity?
2. Was there a trend in your reaction times in steps 1 and 3 of the activity? If so, what did you notice about your reaction time?
3. Did your ability to catch the coin in step 2 improve through the trials?

What Now?

1. How did your nervous system coordinate what was happening in the activity?
2. What ways do you see of improving your reaction time?
3. Would it ever be possible to stop the watch exactly on 5 seconds? Why or why not?

Getting on Your Nerves

As you saw in the Explore Activity, your nervous system allows you to sense and react to objects and events in your environment. Objects or events that cause reactions are called stimuli (stim′ yə lī). Stimuli are perceived by the nervous system. The nervous system consists of **neurons**, or nerve cells, the spinal cord, and the brain. Along the neurons pass the impulses that tell your body what's going on and what to do about it.

There are three basic types of neurons. Motor neurons conduct impulses from the brain or spinal cord to muscles and glands in the body. Sensory neurons receive information and send impulses to the spinal cord or brain.

Interneurons are nerve cells throughout the brain and spinal cord that relay impulses from sensory neurons to motor neurons.

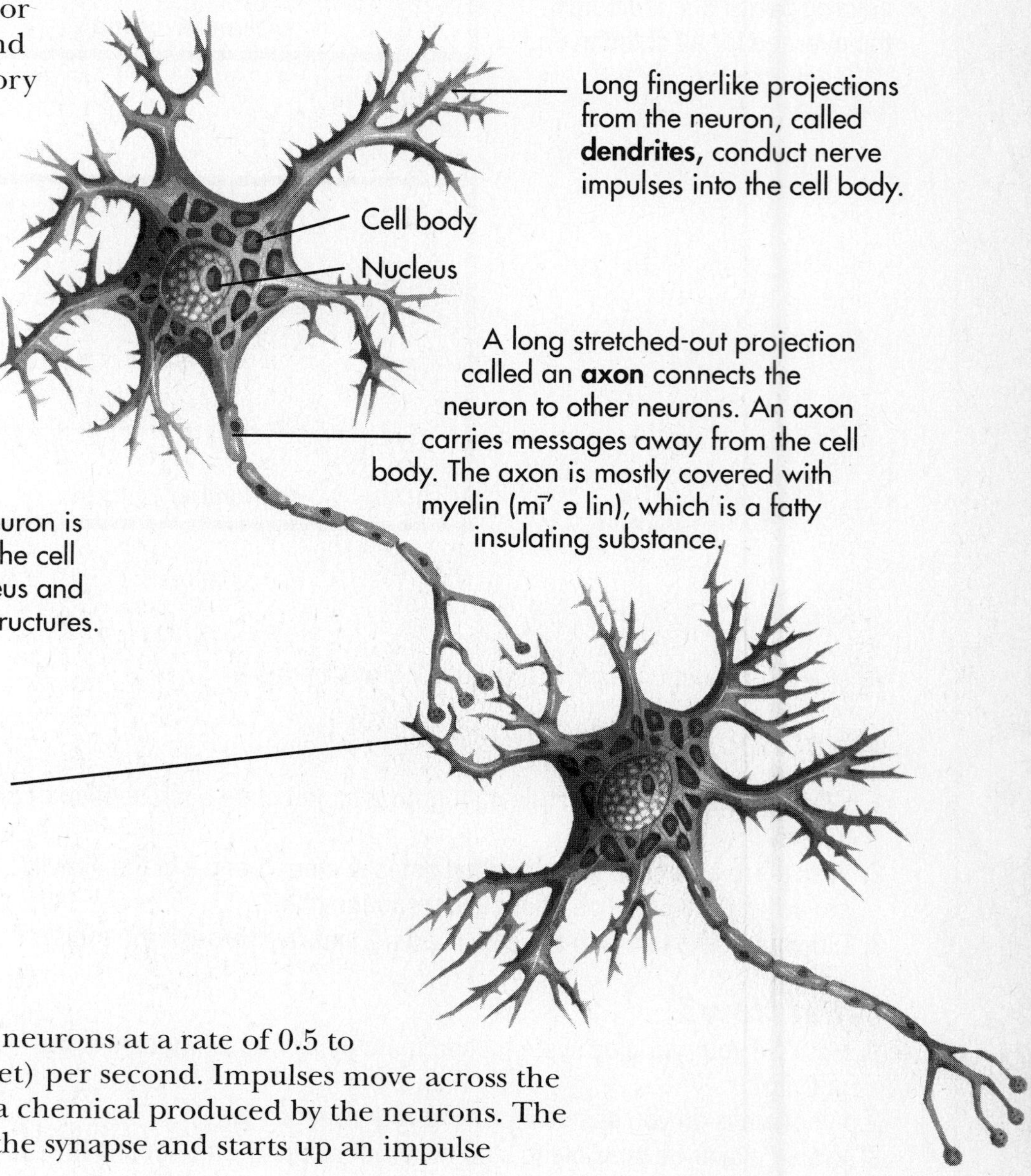

Long fingerlike projections from the neuron, called **dendrites,** conduct nerve impulses into the cell body.

Cell body

Nucleus

A long stretched-out projection called an **axon** connects the neuron to other neurons. An axon carries messages away from the cell body. The axon is mostly covered with myelin (mī′ ə lin), which is a fatty insulating substance.

The main part of a neuron is called the cell body. The cell body contains a nucleus and other important cell structures.

Neurons do not touch each other. There is a small space between them called a **synapse** (sin′ aps).

Impulses move along neurons at a rate of 0.5 to 100 meters (1.6 to 330 feet) per second. Impulses move across the synapse with the help of a chemical produced by the neurons. The chemical diffuses across the synapse and starts up an impulse in the next neuron.

Control Center

The brain is the main organ of the central nervous system where most stimuli are processed. The brain consists of three basic parts, each of which controls different functions—cerebrum, cerebellum and brainstem.

The **cerebrum** (sə rē′ brəm) is the part of the brain in which thinking, sensory perception, memory, and control of voluntary muscles occur. The cerebrum is divided into two hemispheres—the right and the left. Due to a crossover in the nerves, the right side of the brain controls the left side of the body and the left side of the brain controls the right side of the body. Each hemisphere has five lobes, four of which are visible at the surface. Under all the outer lobes, not visible at the surface, is the insula. Evidence suggests the insula may coordinate cerebral functions and may be important in memory. Each lobe has control centers for different body

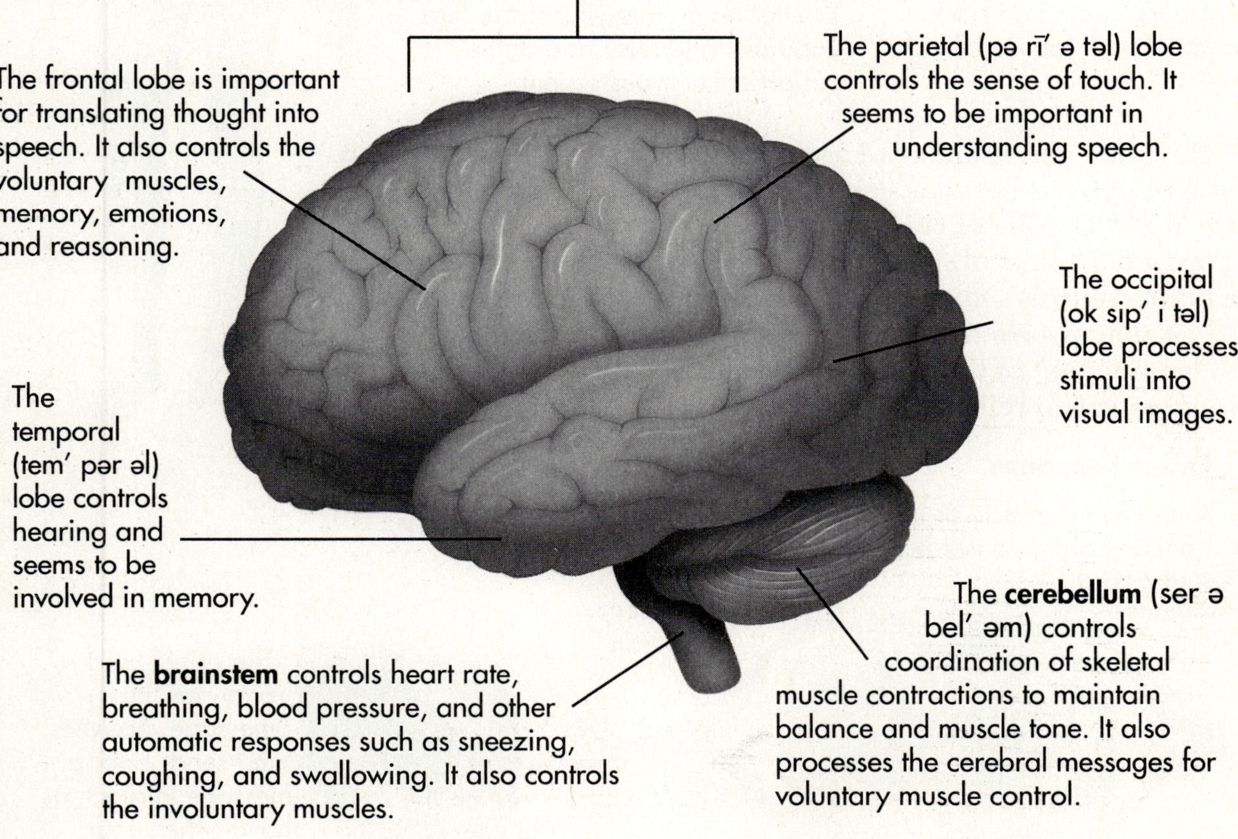

Minds On! You probably aren't aware of what goes on in your cerebellum to maintain your normal body functions. Here's a short activity to help you become aware. First stand up. Put both feet together and stand still. Close your eyes. Are you moving? Are you controlling your motion? Why do you think you are moving? What is controlling the motion? ●

The Interior Network

The second part of the central nervous system is the spinal cord. Most of the nerves that bring messages from your body to your brain branch from the spinal cord. These nerves make up the peripheral nervous system, an interior network that constantly informs your brain and body of where you are and what is happening around you. In addition, it is a network that relays signals from the brain that allow you to do things as simple as walking or as complicated as playing basketball.

Brain

Spinal cord

Nerves

The **spinal cord** is an extension of the brainstem. It is made up of nerve cells that carry impulses from all parts of the body to the brain and back. It is about 43 cm (about 17 in.) long.

The nerves of the spinal cord are protected by your vertebrae. There is a hole in each vertebra through which the nerves of your spinal cord pass. The nerves go into your arms, legs, and trunk and form part of the peripheral nervous system.

The brain and spinal cord are connected to the rest of the body by the peripheral nervous system. The peripheral nervous system is made up of 12 pairs of cranial (from *cranium*, Latin for "skull") nerves and 31 pairs of spinal nerves from the spinal cord. Although the brain controls some reflexes, the spinal cord controls the reflexes that involve the arms and legs.

One function of the peripheral nervous system is to make rapid responses, called reflexes.

Reflex Response

A nerve in a muscle or in the skin or a signal from one of the senses brings a message to the spinal cord.

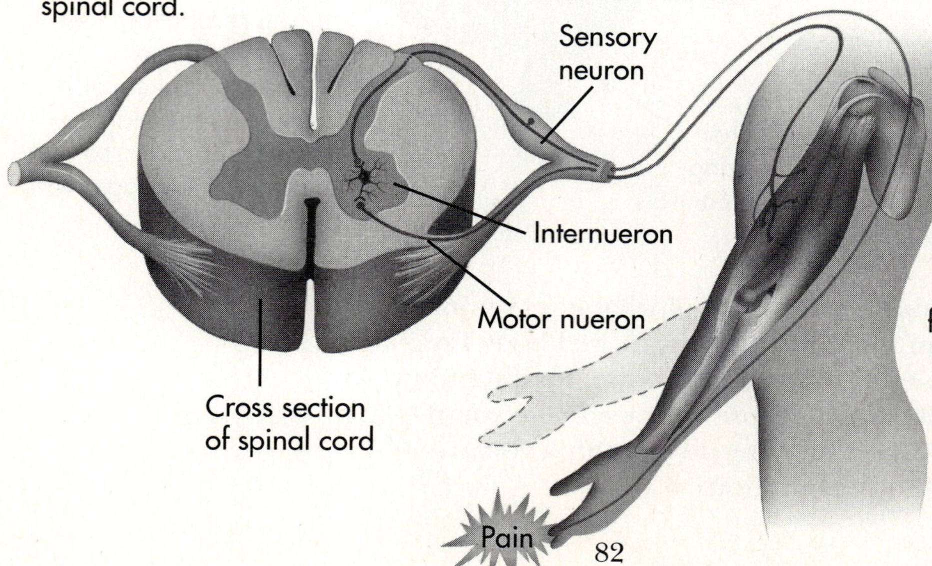

The spinal cord analyzes the message received and sends back a message to react, as shown when your hand quickly moves from a hot surface. The knee-jerk reaction is one reflex that is commonly tested during a doctor's examination.

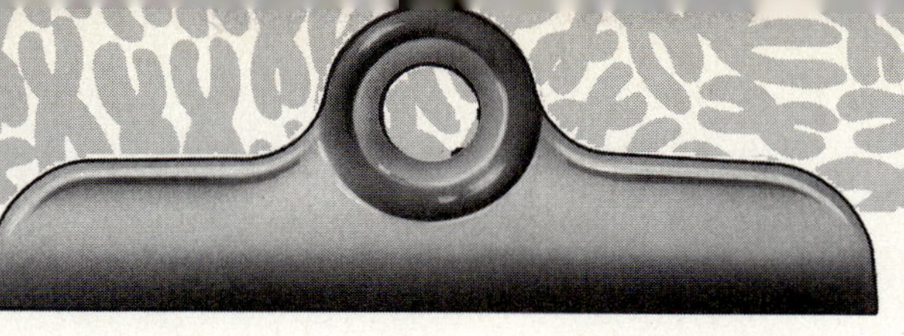

Singular Sensations

You're aware that you receive information from the environment in a variety of ways. Your senses are the main sources for much of your information about the world around you.

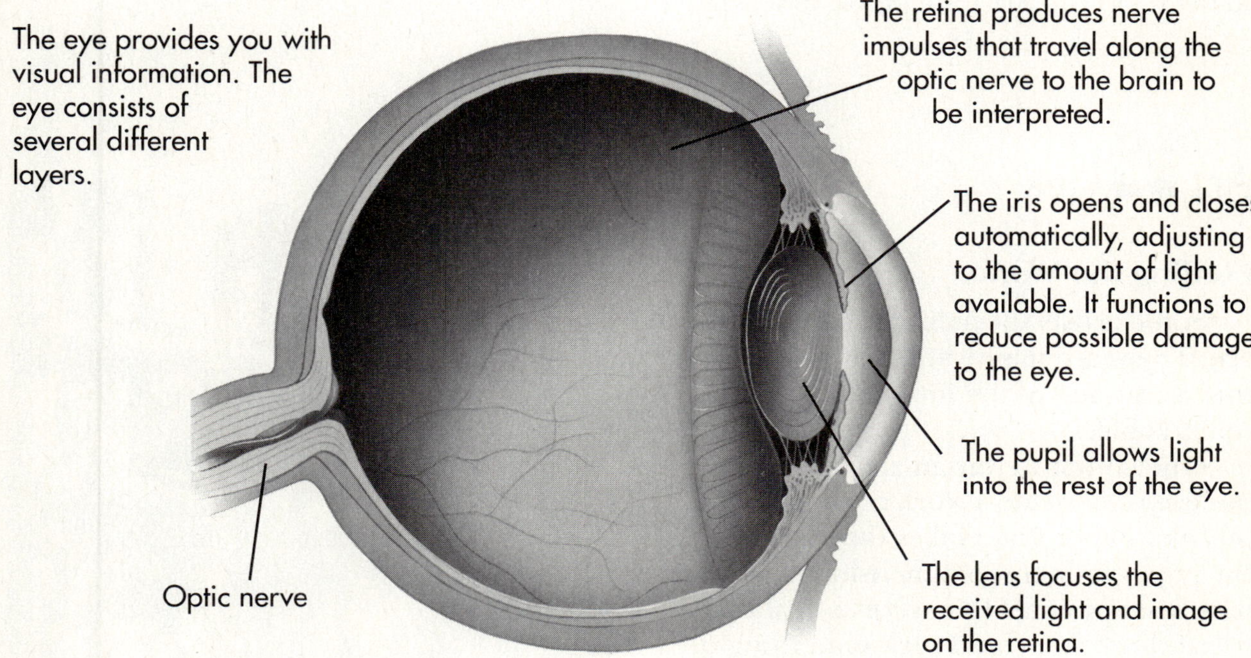

TRY THIS **Activity!**

A Hole in One!

Examine one of the unique features of human vision in this activity.

What You Need

cardboard tube

Close your right eye. With a finger at arm's length, cover a small object at the other side of the room. Now close your left eye and look with your right. Record your observations on another sheet of paper. Hold the cardboard tube to one eye. Keeping both eyes open, hold your hand up about 10 cm from your other eye. Record your observations . What did you see in each of the two steps? Formulate a hypothesis to explain what you saw in each case.

What you observed in the Try This Activity is one result of a characteristic of human vision called stereoscopic vision. With stereoscopic vision each eye sees an object at a slightly different angle. These two images are put together by the brain to make a three-dimensional image. It is necessary to have sight in both eyes to perceive three dimensions.

You experienced one way of fooling your eyes in the Try This Activity. Do the Art Link below to learn about other ways.

Art Link

Fouling the Works

Some artists specialize in tricking the eye. There are many different ways to do this, and some of the images that result are remarkable.

One common way an artist may trick the eye is to create a work of art on a flat piece of paper that makes the objects on the paper look three-dimensional. You may have seen pictures that were so real that you'd think you could pick something off the page. This is the result of one type of illusion.

M.C. Escher was a graphic artist who worked largely in black and white. Some of his pictures show staircases that go only upwards, double-sided staircases, and closely fitting light and dark images and creatures.

Some artists of the early 1900s, including a man named George Seurat, produced paintings in which they did not directly blend paints, but instead painted with very small dots. These small dots are quite close together, and the eye sees the blending of the colors rather than the separate colors. If you examine a cartoon from the Sunday newspaper, you'll see small dots of color similar to those used in paintings by artists like Seurat.

Optical illusions like this one trick the eye.

As a group, choose one of these ways of tricking the eye, or other ways, including optical illusions. Bring in or create your own examples of the technique you research. Present a class report that discusses what the technique is and how it tricks the eye. You may wish to create a poster of your work to show to the rest of the school.

Hearing and Balance

The ear is the sense organ for hearing. The ear is divided into three main sections–the outer ear, the middle ear, and the inner ear. The outer ear consists of the ear that you can see on the outside of a person's head and the auditory canal. Its primary function is to collect sound waves and pass them inward.

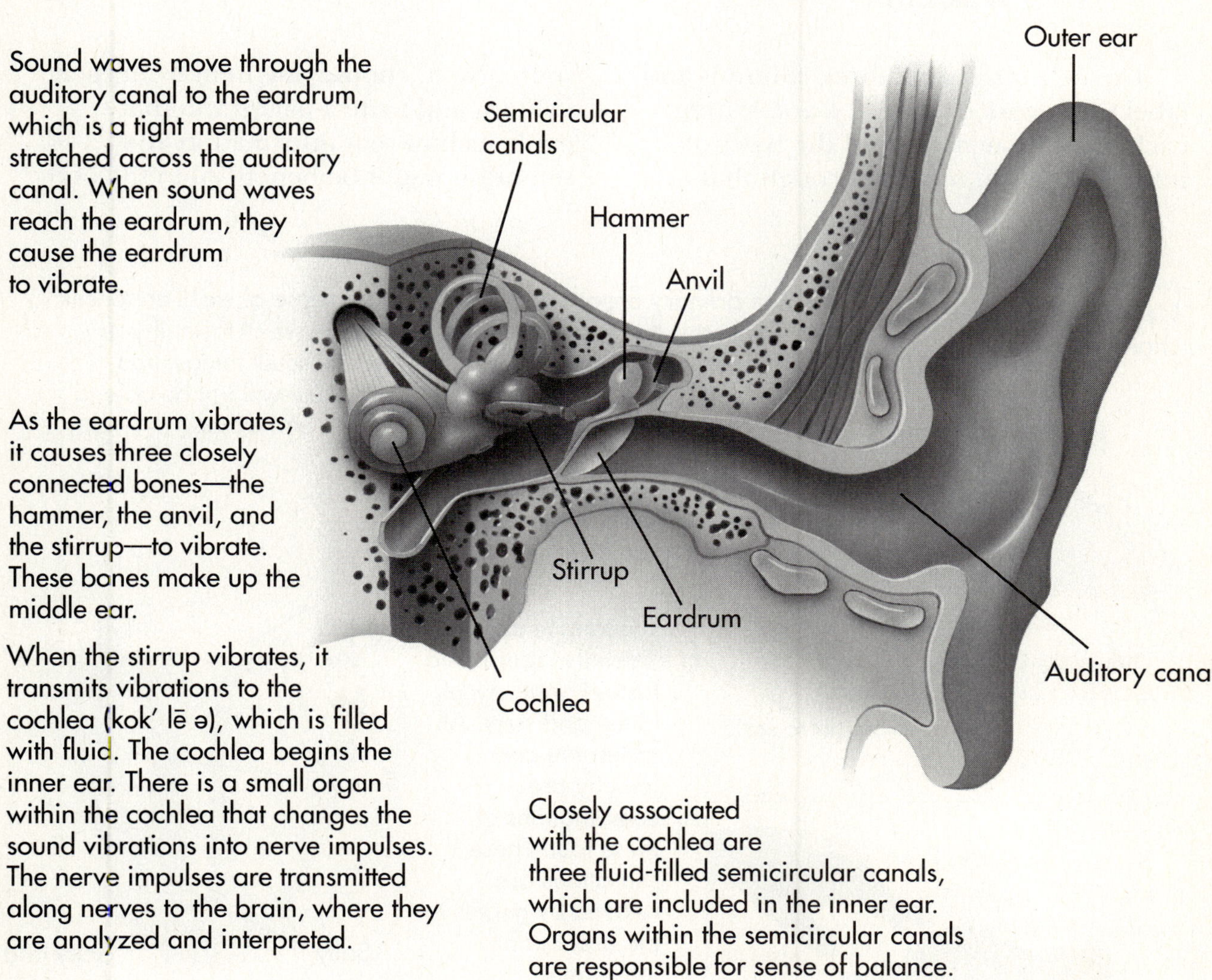

Sound waves move through the auditory canal to the eardrum, which is a tight membrane stretched across the auditory canal. When sound waves reach the eardrum, they cause the eardrum to vibrate.

As the eardrum vibrates, it causes three closely connected bones—the hammer, the anvil, and the stirrup—to vibrate. These bones make up the middle ear.

When the stirrup vibrates, it transmits vibrations to the cochlea (kok' lē ə), which is filled with fluid. The cochlea begins the inner ear. There is a small organ within the cochlea that changes the sound vibrations into nerve impulses. The nerve impulses are transmitted along nerves to the brain, where they are analyzed and interpreted.

Closely associated with the cochlea are three fluid-filled semicircular canals, which are included in the inner ear. Organs within the semicircular canals are responsible for sense of balance.

Minds On! Why is it important to ensure that no damage occurs to the eardrum? How might sound damage the eardrum? Write your answers to these questions and discuss them among your group. Share your answers with the class. As a class, brainstorm precautions you can take to protect your hearing from prolonged exposure to very loud sounds. ●

Smell, Taste, and Touch

Sight and hearing are only two ways to receive information about your environment. Smell, taste, and touch also provide a great deal of information about what is happening in the world around you.

Language Arts Link — The Sense of Sensation

Create a chart with three columns and label them *smell, taste,* and *touch.* Within each column, make lists of the types of information you receive through that sense. Next, choose one item from each column and write a paragraph that explains how that information and sensation might be beneficial to the body.

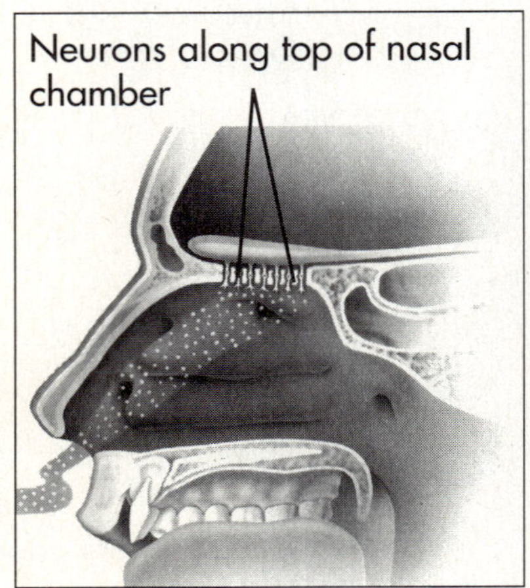

The primary organ used in smelling is the nose. The nose takes in air. About two percent of the air you breathe strikes an organ within the nose. Within this organ, sensory information is turned into nerve impulses that travel to the brain to be interpreted.

Your sense of smell enhances your sense of taste. If you cannot smell things, you probably will not be able to taste them, either.

Illustrations not to scale

The tongue can "identify" four basic tastes—sour, sweet, bitter, and salty. All tastes you can identify are combinations of these four. These sensations are relayed by nerves to the brain to be interpreted.

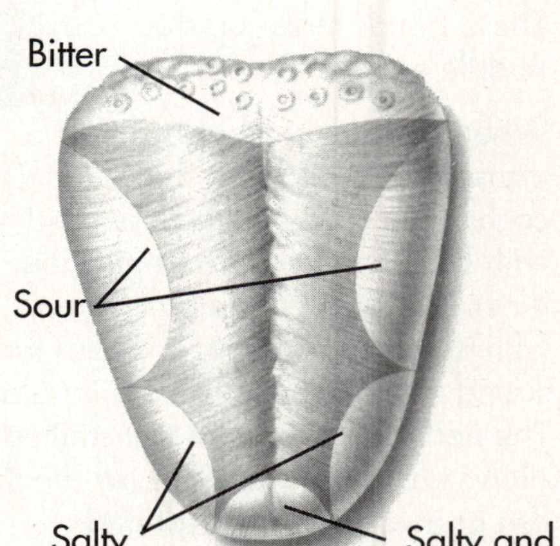

Nerves within the skin can sense light touch, pressure, heat, and pain. They transmit impulses to the brain or spinal column where they are interpreted and reacted to. Each part of the body has a different amount of the brain devoted to it for the sensation of touch.

Diseases and Disorders

There are a great many diseases and disorders associated with the nervous system. Some of these are psychological in nature, while others are physical. Let's focus on the physical.

Epilepsy is a fairly common disorder that may be inherited or may be caused by tumors or head injuries. The symptoms may be as mild as momentarily (for 5–30 seconds) losing contact with reality. Or they may be as severe as extreme convulsions and spasms in which the muscles go rigid for seconds at a time. The person experiencing this extreme form of epilepsy may be unconscious for some period of time.

Meningitis is a bacterial or viral infection of the lining of the spinal cord or brain. Some types of meningitis may result in only a headache and a fever. Others may cause convulsions, coma, and death. Meningitis that affects the brain and its lining is called encephalitis. Encephalitis is commonly spread by bites from infected mosquitoes.

Technology can provide devices that assist persons with disabilities in a variety of activities.

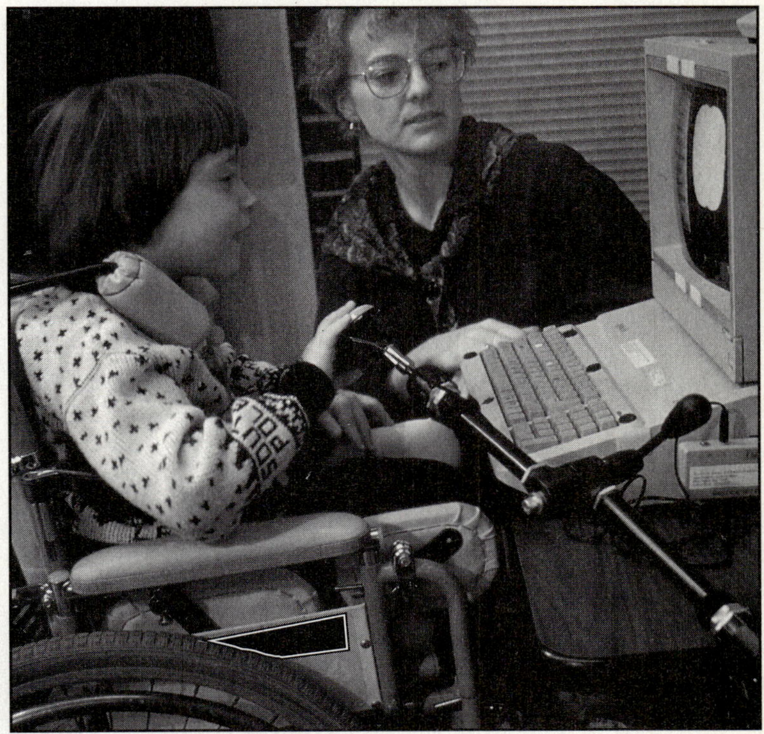

Multiple sclerosis is a disease of the nervous system in which the myelin sheaths that surround the nerves begin to degenerate and harden. This makes it more difficult for nerve impulses to be transmitted. As a result, a person with multiple sclerosis may find it more and more difficult to perform ordinary tasks. There is no cure for this disease.

Choose one of these diseases or disorders, or research others that involve the nervous system. Find out the symptoms, how it is treated, and whether or not it can be prevented or cured. Your group is going to be the "panel of experts" on the disease or disorder. After a brief presentation, ask for questions from the class. The class will be groups of medical reporters who want to inform the public of any new developments. Be ready to explain what you have learned in terms that everyone can understand. Your object is to convey information to as many people as possible.

Focus On Technology

Be Careful With That Brain

The brain is one of the most important and most delicate organs of the body. Sometimes things go wrong with the brain or with other parts of the body associated with the brain, and it is necessary to do surgery. In one recent case, doctors had to alter the shape of a little girl's skull because the bones of the skull had fused together too early, and the growth of the brain was distorting the skull. Since it involved operating on bones that surrounded the brain, the surgery was very delicate. The doctors first practiced the surgery many times on a computer before they operated on the girl. As a result the actual surgery went smoothly, and the little girl's skull was reshaped with no damage to her brain.

Some infants and young children have extremely bad cases of epilepsy that cause convulsions and unconsciousness almost from birth. Doctors knew from their experience with adults and from test results that part of the problem was the connection between the right half and the left half of the brain. One former treatment for epilepsy in adults was to cut the network of nerves that linked the two hemispheres of the cerebrum. However, recently one doctor concluded that if half of the brain were removed or completely disconnected, the problem would be resolved. This operation has been performed on a number of children. The children survived the operation and seem to have recovered from it well. There are some side effects, such as loss of sight in one eye and loss of the full use of one arm.

Only through a full understanding of how technology is best used in medicine can we hope to advance. If techniques are not analyzed and questioned, possible useful results of a procedure may be overlooked. As a group, debate the pros and cons of using a computer to practice or analyze surgical techniques. Or debate the pros and cons of performing radical operations like removing half a brain. Make a list of what you see as the advantages and another list of problems and questions you have about the techniques. If possible, research the problems you have listed, or speak with a doctor regarding some of these problems.

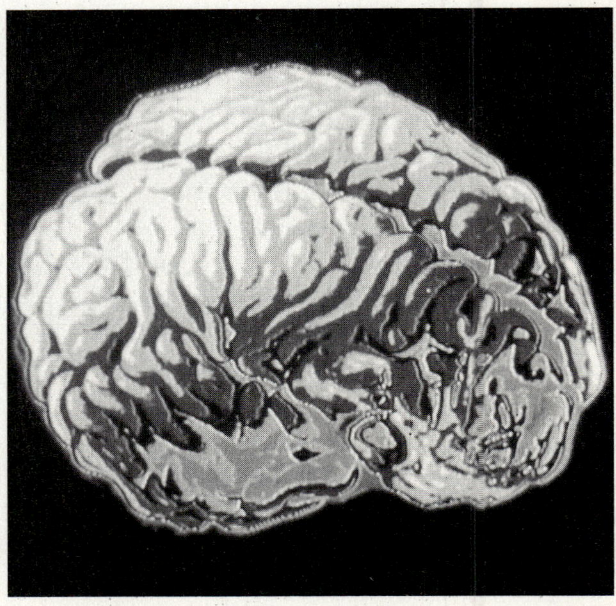

Computer-generated image of a human brain

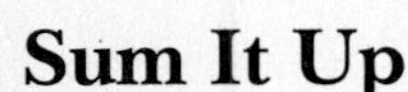

Sum It Up

The central nervous system is made up of the brain and spinal cord. The peripheral nervous system is made up of the cranial nerves and the spinal nerves. The brain controls the functioning of the whole body and helps coordinate responses to the environment. The senses provide the brain with information from the environment. Using information from both inside and outside the body, the brain is able to regulate and coordinate body functions so that you can function in the world around you.

Using Vocabulary

Use the vocabulary words to describe the transmission of messages from the brain to the heart and muscles in the leg.

axon	cerebrum	spinal cord
brainstem	dendrites	synapse
cerebellum	neurons	

Critical Thinking

1. Explain how reflexes help to protect the body.

2. Compare the functions of the cerebrum, cerebellum, and brainstem.

3. If something happened to someone's cerebrum, is it possible that that person's heart could continue to beat normally? Why or why not?

4. Why is it important that nerve cells are insulated from their surroundings?

5. What might be a result of losing your sense of touch?

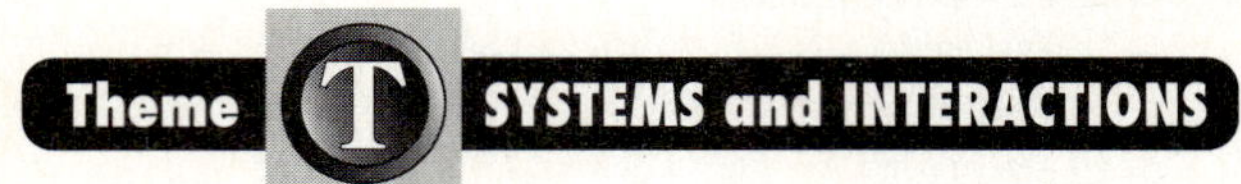

Life Choices

In this unit, you've studied how body systems work together to perform specific functions that are necessary to maintain life. You've also seen that there are ways to help body systems continue to function well throughout life. Now, put together what you've learned in this unit and make a plan that will keep your body functioning at top form.

Start by analyzing your daily activities. Look at your meal plan from Lesson 1 and compare it to your current eating habits. Did you try your meal plan for a week? What foods did you find most difficult to give up or to eat as part of your healthy diet? What suggestions do you have to improve your own diet?

Take the information you learned in Lessons 2 and 3 about the healthful benefits of exercise. Think back over the time you've spent studying this unit and try to make up a list of your general activities for a day. How does it compare to a healthful exercise routine? What can you do to improve it?

Finally, use the information about causes of disease and injury that you studied in Lessons 3, 4, and 5. What choices can you make to reduce your chances of getting some of these diseases? What can you do in each of your daily activities to reduce the risk of injury?

How long has it been since you've tried a vegetable that was new to you?

Now consider some things that you do on a regular basis but not every day. For instance, to make certain that your body is functioning normally, you probably visit a doctor once a year for a checkup. List the tests that the doctor does during this visit. How do these tests help check body systems?

Also, there may be times when you need to seek medical attention. Make two lists on a separate sheet of paper. In the first list, place diseases and conditions such as a very high fever for which you should see a doctor. In the second list, place diseases and conditions that you do not need to see a doctor to treat.

Write a short paper that summarizes your conclusions about choices that lead to proper body functioning. In your paper, state those things that you already do and those you would like to start doing in the near future.

The human body is well equipped to find, obtain, and use the necessities of life—food, water, oxygen, protection, and companionship. Understanding how body systems work together to maintain these functions is the first step in leading a healthy life.

Swimming is only one of many ways to get exercise and have fun.

GLOSSARY / INDEX

Use the pronunciation key below to help you decode, or read, the pronunciations.

Pronunciation Key

a	at, bad
ā	ape, pain, day, break
ä	father, car, heart
âr	care, pair, bear, their, where
e	end, pet, said, heaven, friend
ē	equal, me, feet, team, piece, key
i	it, big, English, hymn
ī	ice, fine, lie, my
îr	ear, deer, here, pierce
o	odd, hot, watch
ō	old, oat, toe, low
ô	coffee, all, taught, law, fought
ôr	order, fork, horse, story, pour
oi	oil, toy
ou	out, now
u	up, mud, love, double
ū	use, mule, cue, feud, few
u	rule, true, food
u̇	put, wood, should
ûr	burn, hurry, term, bird, word, courage
ə	about, taken, pencil, lemon, circus
b	bat, above, job
ch	chin, such, match
d	dear, soda, bad
f	five, defend, leaf, off, cough, elephant
g	game, ago, fog, egg
h	hat, ahead
hw	white, whether, which
j	joke, enjoy, gem, page, edge
k	kite, bakery, seek, tack, cat
l	lid, sailor, feel, ball, allow
m	man, family, dream
n	not, final, pan, knife
ng	long, singer, pink
p	pail, repair, soap, happy
r	ride, parent, wear, more, marry
s	sit, aside, pets, cent, pass
sh	shoe, washer, fish, mission, nation
t	tag, pretend, fat, button, dressed
th	thin, panther, both
<u>th</u>	this, mother, smooth
v	very, favor, wave
w	wet, weather, reward
y	yes, onion
z	zoo, lazy, jazz, rose, dogs, houses
zh	vision, treasure, seizure

alveoli (al vē′ ə lī)**:** groups of thin-walled sacs in the lung where oxygen and carbon dioxide are exchanged, **p. 40**

aorta (ā ôr′ tə)**:** the main artery of the body through which blood leaves the heart, **p. 41**

atrium: either one of the two upper chambers of the heart, **p. 41**

axon: a long extension of a neuron that carries messages away from the cell body, **p. 80**

brainstem: the part of the brain responsible for controlling heart rate, breathing, and other automatic responses of the body, **p. 81**

capillaries (kap′ ə ler′ ēz)**:** microscopic blood vessels that connect arteries and veins through which oxygen, nutrients, and wastes are exchanged, **p. 42**

carbohydrates (kar′ bō hī′ drāts)**:** nutrients composed of carbon, hydrogen, and oxygen that provide most of the body's energy and include simple sugars and starches, **p. 12**

cardiac muscle: involuntary muscle tissue found only in the heart, **p. 32**

cerebellum (ser ə bel′ əm): the part of the brain responsible for maintaining posture, coordinating movements, and processing the impulses of the cerebrum to the voluntary muscles, **p. 81**

cerebrum (sə rē′ brəm): the part of the brain responsible for interpreting impulses from the senses, controlling voluntary muscles, and thinking, **p. 81**

childhood: the period of life from the end of infancy (age 1) to puberty, **p. 71**

compact bone: the outer, dense, hard layer of bone, **p. 27**

dendrite: extension of a neuron that receives messages and sends them to the cell body, **p. 80**

endocrine system: a system of glands that secrete chemicals called hormones directly into the bloodstream, **p. 72**

enzymes (en′ zīmz): chemicals formed in living cells that speed chemical reactions without being changed themselves, **p. 16**

fats: one of three main types of nutrients; composed of carbon, hydrogen, and oxygen, **p. 12**

haversian (hə ver′ zhən) **system:** a system of interconnecting hollow areas within the compact bone that supplies the bone with blood vessels and nerves, **p. 27**

hemoglobin (hē′ mə glō′ bin): a chemical in red blood cells to which oxygen attaches to be carried throughout the body; gives blood its red color, **p. 43**

immune system: a body system composed of tissues that fight infection and disease in the body, **p. 57**

immunity: the body's ability to resist or fight off infection, **p. 60**

infancy: the period of life between birth and age 1, **p.71**

integumentary (in teg′ yə men′ tə rē) **system:** the system consisting of hair, nails, and skin that serves as an outer protective covering for the body, **p.54**

large intestine: the part of the digestive system through which waste materials pass and in which water is absorbed, **p. 18**

lymphatic (lim fat′ ik) **system:** a system consisting of vessels that circulate lymph (a fluid derived from blood plasma) throughout the body; collects lymph from body tissues, filters it, and returns it to the blood, **p. 56**

neuron: one type of nerve cell, **p. 80**

ovaries: the organs of the female reproductive system in which eggs mature, **p. 69**

phagocytes (fag′ ə sīts): cells that function as part of the immune system by consuming foreign particles in the bloodstream and lymph, **p. 57**

platelets: the components of blood responsible for clotting, **p. 43**

protein: one of three main types of nutrients; composed of carbon, hydrogen, oxygen, and nitrogen, used throughout the body for growth and to replace and repair cells, **p. 12**

puberty: the period of life when the reproductive system matures to the point where a person is capable of reproducing, **p. 71**

red marrow: material found within the spongy bone that produces red blood cells, **p. 27**

reproductive system: the system responsible for the continuation of a species, **p. 68**

respiration: the process of turning food into energy, **p. 31**

skeletal muscle: voluntary muscle tissue attached to bones that makes bones move, **p. 32**

small intestine: part of the digestive system between the stomach and the large intestine where most digestion and absorption of food normally occurs, **p. 16**

smooth muscle: involuntary muscle tissue that forms the walls of the stomach, intestines, uterus, and blood vessels, **p. 32**

spinal cord: a long cord of nerve tissue that extends from the brain through the upper two-thirds of the backbone; an extension of the brainstem made up of bundles of neurons that carry impulses from all parts of the body to the brain and from the brain to all parts of the body, **p. 82**

spongy bone: the inner layer of bone (found toward the ends of bones), in which red and yellow marrow exist, **p. 27**

stomach: the part of the digestive system in which food is churned with digestive secretions before being released into the small intestine, **p. 16**

synapse (sin′ aps): the space between two nerve cells across which impulses can travel, **p. 80**

testes: the organs of the male reproductive system in which sperm mature, **p. 68**

vena cava (vē′ nə kā′ və): one of two large veins of the circulatory system through which oxygen-poor blood returns to the heart from the body, **p. 41**

ventricle: either one of two lower chambers of the heart, from which the blood is pushed out into other parts of the circulatory system, **p. 41**

villi (vil′ ī): small fingerlike projections of the small intestine through which nutrients are absorbed into the bloodstream, **p. 18**